ELECTRONIC CIRCUIT ANALYSIS USING LTSPICE XVII SIMULATOR

ELECTRONIC CIRCUIT ANALYSIS USING LTSPICE XVII SIMULATOR

A Practical Guide for Beginners

Pooja Mohindru
Pankaj Mohindru

CRC Press is an imprint of the
Taylor & Francis Group, an **informa** business

First edition published 2022
by CRC Press
6000 Broken Sound Parkway NW, Suite 300, Boca Raton, FL 33487-2742

and by CRC Press
2 Park Square, Milton Park, Abingdon, Oxon, OX14 4RN

First edition published by CRC Press 2022

CRC Press is an imprint of Taylor & Francis Group, LLC

Library of Congress Cataloging-in-Publication Data
A catalog record has been requested for this book

ISBN: 978-1-032-04076-9 (hbk)
ISBN: 978-1-032-05847-4 (pbk)
ISBN: 978-1-003-19948-9 (ebk)

Typeset in Times
by MPS Limited, Dehradun

Contents

Preface

LTspice XVII is a high-performance and easy-to-use freely available SPICE-based circuit simulator for analyzing electronic circuits. Before a hardware implementation is done, analyzing electronic circuits' behavior using the simulation software is essential to minimize the manual efforts in doing theoretical calculations, soldering, de-soldering, etc. Hardware implementation should be executed only when accurate simulation results are obtained to save time and money. The purpose of this book is to facilitate graduates/postgraduates, industry professionals, researchers, academicians, etc. in acquiring knowledge for using the LTspice XVII schematics and support programs effortlessly and effectively. It is a sincere attempt by both authors to cover all the topics comprehensively and interestingly. The authors decisively believe that the book will prove to be extremely valuable, productive, and could hone the requisite skills in simulation analysis using the schematic capture.

Abstract

LTspice XVII is a freely available full-featured SPICE-based circuit simulator that is developed by Analog devices (originally by Linear Technology). It includes the schematic entry/capture for drafting a circuit, the waveform viewer for probing and graphical plotting, the library of component models, and easy-to-use advanced features to analyze the performance of the circuit design. SPICE (Simulation Program with Integrated Circuit Emphasis) computer software is a widely circulated general-purpose simulator for modeling and analyzing electronic (specifically analog) circuits before implementing them onto an integrated circuit, and thus cannot be separated from the analog IC design process. SPICE simulators are the only way to check the circuits' behavior formerly to integration onto a chip. LTspice XVII enables the users to iterate their designs in minimal time by doing successive simulations so that we can explore the circuit limitations and performance boundaries of the design with slight adjustments. Further, the SPICE simulation allows interactive measurements of currents, voltages, and power that are almost not possible to compute theoretically or any other way. LTspice XVII has an integrated schematic editor that includes a wide-ranging library of electronic and electrical component models for resistors, capacitors, inductors, diodes, OP-Amps, transistors (BJTs, FETs, MOSFETs), ICs, switches, independent and dependent sources, etc. A circuit diagram is drafted graphically as a list of the components connected on the new schematic using the simulator schematic capture program (node/net numbers are assigned automatically to the connection points). A digital (simple logic gate) simulation capability is its added feature. Most importantly, it has no constraints on the number of nodes, component models, sub-circuits, or measurements. Also, voltage and current probes can be added easily to the circuit schematic. The simulation software is extremely useful in several areas such as electronics (power electronics, radio frequency electronics, audio electronics, and digital electronics), electrical science (simple and complex circuits' analysis), mechanical systems, mechanics, physics, and other disciplines. Designing electronic circuits using the simulation software is essential so that simulation of circuits can be performed and their behavior can be analyzed before a hardware implementation is done to minimize the manual efforts in doing theoretical calculations. Hardware implementation must always be executed only when accurate simulation results are obtained so that our time and money can be protected. All kinds of circuits can be defined and simulated in LTspice XVII either by drawing them on the Schematics or writing coded netlist files. The book is intended to explain simulations with schematic files only. The book is about using LTspice XVII (the latest version) for analog and digital circuit simulations. It covers simulation and analysis of various electrical and electronic circuits used frequently in real-time applications to provide imperative insights into their behavior. The book assists readers in learning analog electronics practically by simulating various circuits with different parameter values of the components (to fit the simulation needs) and comparing the results. It enables the users to get rid of all their fears regarding basic electrical circuit analysis. The approach followed by the authors is to teach the much-needed theory of various electronics and electrical-based circuits along with their testing and software simulations. The aim is that the readers can utilize the simulation outputs (in tabular and graphical forms) to gain a deeper insight into the circuits. The book represents the use of LTspice XVII for Windows as a general-purpose schematic driven program with an integrated SPICE simulator. The schematic capture allows the user to draw a new circuit (or begin with an already drafted circuit) and observes its operation in the simulator for different parametric values of the component models so that the desired circuit behavior can be achieved through iteration. The graphical schematic is finally converted to a textual SPICE netlist and passed to the simulator for execution. An imported netlist can also be run directly without having a schematic. The contents of the book carry significance to both professionals and researchers in various fields. The book can be employed used as an important resource for engineering coursework and professional training. The book is broadly divided into two stages, an introduction explaining the basic functionality of the software followed by simple examples taken for analysis along with a discussion of issues that occur in the practical implementation of those electronic circuit examples and how to resolve them. Our book enables the users to diagnose and treat most of the problems they are likely to come across in day-to-day circuit analysis with the software. The first chapter involves a comprehensive introduction about using all the available LTspice XVII simulator menus, action toolbars, extensive library of component labels and models of semiconductor devices (with many parameters), circuit constructing features, waveform viewing features, interface control options, shortcut keys related information, simulation and analysis types, probing outputs and making measurements, editing sources (to produce more customized simulations), etc. The other chapters covered in the book include the basics of designing electronic, electrical, and communication circuits using the software and observing the output behavior for different parametric values. In all the chapters, the performance of the tool is presented in detail. Also, the simulation process for various circuits is discussed step-by-step for more clarity and understanding to the readers. To conclude, the Schematics include SPICE (a simulation engine) and PROBE (a plotting quality). The book is all about learning to use Schematics and its support programs.

Acknowledgments

We owe a vast debt of gratitude to the almighty for making it possible for us to write the book. We express sincere gratitude to God for inculcating strength, knowledge, intellectual insight, temperament, energy, and passion in us.

We are extremely obliged to Dr. S.S. Bhatia and Dr. Rajesh Khanna, Thapar University, Patiala for their wonderful guidance and motivation. We deeply acknowledge the assistance received from Dr. Manjit Singh Patterh and Dr. Manjeet Singh Bhamrah, Punjabi University. We are also immensely grateful to all our colleagues.

Last, but not the least, we acknowledge all the family members for tolerating us and providing untiring support, love, and encouragement. We especially thank Mrs. **Raj Mehta** and Manasvi for being our unrelenting source of inspiration. An enormous recognition to all those who always ensure us that good times keep flowing.

The author Dr. Pooja takes pride in dedicating the book to her incredibly hard-working and affectionate grandparents, **Sh. Lal Chand Mehta** and **Smt. Hardevi Mehta.** The author is deeply indebted to them for showering their blessings from heaven.

Smt. Hardevi Mehta

Sh. Lal Chand Mehta

Authors

Dr. Pooja Mohindru obtained her Master of Engineering as well as Ph.D. degree in Electronics and Communication Engineering from Thapar University, Patiala, India. She is currently working in the Department of Electronics and Communication Engineering at Punjabi University, India. Together with 16 years of teaching experience, her research focuses on digital signal processing, communication, analog electronics, digital electronics, and microelectronics. Her publications comprise over 25 research articles in various reputable international journals (also include SCI Indexed Journals). She has also published two books titled *Introduction to Fractional Fourier Transform* and *MATLAB and SIMULINK (A Basic Understanding for Engineers).*

Dr. Pankaj Mohindru received his Master of Engineering degree with distinction in Electronics from Thapar University, Patiala, India, in 2005. He has been awarded a Ph.D. by Punjabi University, Patiala, India in the year 2011. Presently, he is working as a senior faculty in the Electronics and Communication Engineering Department, Punjabi University, Patiala. His research area includes fuzzy logic, measurement science and techniques and research methodology, analog electronics, etc. His publications include 35 research papers in reputable national and international journals. He has also published one book titled *MATLAB and SIMULINK (A Basic Understanding for Engineers).*

1 Introducing LTspice XVII Circuit Simulator

1.1 INTRODUCTION

LTspice is a high-performance and easy-to-use circuit simulator based on the SPICE (Simulation Program with Integrated Circuit Emphasis) software. Initially, the general-purpose simulator SPICE was developed at the Electronics Research Laboratory of the University of California, Berkeley (1975) for modeling, analyzing, and displaying electronic (specifically analog) circuits' behavior. LTspice XVII (the latest version) software package is an excellent schematic-driven circuit simulation program that uses SPICE algorithms to run industry-standard semiconductor and behavioral model simulations very fast. The schematic is a user-friendly interface that allows the user to build circuits (with standard component symbols or models) that can be seen on the screen directly so that they can easily understand them and simulate them without employing the circuit defining SPICE data statements with a specific format or syntax. **The schematic capture program transforms the schematic diagram into a netlist which can be easily understood by SPICE. The generated netlist for a particular circuit can be further saved and used in a subsequent analysis of the same circuit.**

A digital (simple logic gate) simulation capability is its added feature. A set of wide-ranging enhancements is included, such as parallel processing, dynamic assembly, and object code generation in the SPARSE matrix solver to make LTspice XVII (an enhanced SPICE type analog electronic circuit simulator) the industry unmatched analog simulator.

LTspice XVII has a wide-ranging library of electronic and electrical component models for resistors, capacitors, inductors, diodes, OP-Amps, and transistors which can be placed and connected on the schematic with a mouse to build the desired circuit diagram. The user needs only to enter specific numeric values (variables/expressions should be entered using the simulator valid syntax) in a list of the parameter values provided for the device models stored internally so that the models can perform their intended operation. It provides a waveform viewer for visualizing simulation results. All these features of the simulator make the use of a SPICE program much easier and accessible task for the users. Also, LTspice XVII can include SPICE models of complex analog components like op-amps, voltage regulators, and timers into the circuit.

The schematic is a graphical interface to the circuit netlist generation, that is, the simulator interprets the circuit entered graphically on the schematic by a text netlist generated in the background. The netlist consists of a list of the circuit elements and their nodes, model definitions, and other SPICE commands. LTspice XVII saves schematic drafts with a file name extension of .asc. When a circuit drawn on the schematic is simulated, the netlist information is extracted from the schematic graphical information to a file with the same name as the schematic but with a file extension of .net.

Thus, it is a widely circulated and used simulation software in several areas, which include radio frequency electronics, power electronics, audio electronics, digital electronics, mechanics, physics, and other disciplines. Even though it is freely available, the simulator is not constrained to have bounded capabilities (which imply there is no limit to nodes, components, and sub-circuits), and thus outperforms many simulation solutions in the market.

1.1.1 Need for Electronic Circuits Computer Simulator

It is extensively significant to use SPICE simulators for testing before constructing a new circuit in hardware because simulating and analyzing the designed circuit using the software is easier, time-saving, and less expensive.

Testing with SPICE simulators reduces the wastage of hardware resources. This is because hardware implementation is executed only when accurate simulation results are obtained. On the other hand, when a circuit is tested after fabricating it on a printed circuit board (PCB) and desired results are not achieved, components can be damaged while taking them out of the real circuits. Also, re-soldering may affect the performance of the circuit.

Again, simulation using the software takes much lesser time as is consumed in analyzing the circuits' responses theoretically.

It is pretty easy to draw and configure a circuit in the software and probing various points to examine the waveforms.

The users can employ the software as a starting point for the development and modification of their designs.

It is possible to test a new thought/idea without a soldering iron.

An existing idea can be easily modified and tested without a real PCB.

1.1.2 Advantages

i. It is free and provides unlimited circuit sizes.
ii. It allows adding new models and changing the simulator's behavior easily.
iii. It saves time by eliminating the need of solving the values of voltage, current, and power of any component in a circuit manually.
iv. It allows changing the values of components in a circuit so that the circuit can be tested easily before selecting values of components to construct the circuit in hardware.
v. Tells the feasibility of the circuit.
vi. LTspice enables the users to iterate their designs in minimal time by doing successive simulations so that we can explore the circuit limitations and performance boundaries of the design with slight adjustments. Further, the SPICE simulation allows interactive measurements of currents, voltages,

and power that are almost not possible to compute theoretically or any other way.

vii. SPICE simulators are the only way to check the circuits' behavior prior to integration onto a chip.

As compared to LTspice XVII, PSpice and Electronic Workbench are better at mixed analog/digital circuits.

1.1.3 Hardware Necessities

LTspice XVII runs on 32-bit or 64-bit editions of Windows 7, 8, 8.1, 10, Windows XP, and MacOS 10.9+. As a simulation may generate many megabytes of data in a few minutes, it is endorsed to have free hard disk space (>10 GB) and a large amount of RAM (>1 GB). Fundamentally, the program can run on any PC with Windows 98 or above, but the simulation may not finish if there is not enough hard disk space.

1.2 LTspice XVII MAIN INTERFACE TO GET STARTED

LTspice XVII is a freeware software for circuit design and simulation using various key circuit simulation and analysis types such as transient (time-dependent), noise, AC, DC, DC transfer function, DC operating point, parametric, temperature sweep, as well as Fourier analysis. It provides the schematic capture (to enter a schematic for an electronic circuit), the waveform viewer (to display the results of a simulation), extensive library of passive devices, unlimited number of nodes, integrated voltage and current probes, FFT function, far-reaching help function, and automatic update. The simulator also allows entering different mathematical expressions in the traced output using the valid functions. It also permits the computation of heat dissipation of the components and the generation of efficiency reports. It permits fast simulation of switching mode power supplies and also provides advanced analysis and simulation options. It contains over 200 Op-Amp models, transistor models, MOSFET models, etc. LTspice XVII does not generate PCB layouts, but netlists can be exported to PCB layout software. Thus, LTspice XVII assists in visualizing a circuits' responses to arbitrary inputs before they are built.

An LTspice XVII software simulator can be driven by adopting either of the following two modes.

A circuit design and testing process using LTspice XVII can be done in two ways as follows:

a. Using the advanced program called Schematics, which allows the user to draw the circuit diagram and assign the element values through user-friendly dialog boxes. The schematic editor is used as a general-purpose schematic driven capture/SPICE to draw a circuit graphically (netlists are extracted from the drafted schematic diagram) and observe its operation in the simulator. A circuit can be tested and simulated until the drafted circuit shows the desired output performance. As a SPICE simulator only understands netlists, the graphical schematic drafted is finally transformed to SPICE-approved text netlist files.

b. By typing hand-written netlist files (textual inputs having data statements with a specific format, no schematic) in exact order for the simulator to have a list of detailed circuit information (names and values of components and sources and how they are connected). A circuit diagram may be entered but its significance is merely to make a simulation easier for the user to understand. In this method, it is difficult to recognize mistakes.

To conclude, the book is intended to represent the use of LTspice XVII for Windows as a general-purpose schematic driven program with an integrated SPICE simulator. The schematic capture allows the user to draw a new circuit (or begin with an already drafted circuit) and observes its operation in the simulator for different parametric values of the component models so that the desired circuit behavior can be achieved through iteration. The graphical schematic is finally converted to a textual SPICE netlist and passed on to the simulator for execution. An imported netlist can also be run directly without having a schematic.

1.2.1 Update LTspice XVII Using the Sync Release

To keep the software updated with the latest models, software, and examples, it is required to synchronize the release of LTspice XVII using the option **Sync Release** that is available under the **LTspice Tools** menu (Fig. 1.1).

It is always significant to use the Sync Release every time we use the software to have the latest models loaded.

1.2.2 The Desktop Start-Up Screen and Schematic Editor Window

LTspice XVII software is available for various operating systems (Windows 7, 8 and 10, Mac OS X 10.9+, Windows XP, etc.) and the downloadable installation files for it are freely available at analog.com/LTspice. It employs LTspice as its circuit simulation engine that is one of many modern derivatives of the SPICE program originally developed at UC Berkeley in the 1970s.

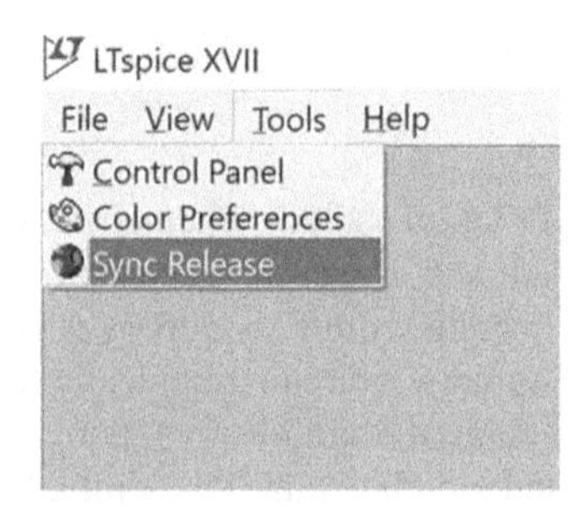

FIGURE 1.1 The Sync Release Button for Updating the Software.

- **Downloading and Installation**

Download an installation file of LTspice XVII, which runs on Windows from the URL *analog.com/LTspice* (earlier http://www.linear.com/designtools/software/). **Copy and paste the URL into a web browser and** select the option Download for Windows 7, 8, and 10. This prompts the user to save the executable file LTspiceXVII.exe. After the file is downloaded, the user can select it to run the installation program, agreeing to the standard Microsoft query about installing a file downloaded from the internet. After maintaining the default installation directory as C:\Program Files\LTC\LTspice XVII, accept the license agreement, and then select the option Install Now to install LTspice XVII (SPICE-based software).

- **How To Open the Start-Up Window**

Using the LTspice XVII Icon for Windows Platform: After completing the installation of LTspice XVII that runs on the **Windows** operating system, go to the Start icon on a computer desktop, and then from its programs list, select the LTspice XVII shortcut icon .

This opens up an initial LTspice XVII start-up window containing all the functionalities of the software as shown in Fig. 1.2. Although the menus and toolbar icons/buttons can be seen at the top of the desktop window, only limited icons are colored and in an active state which implies that the uncolored icons are not active at present. To make most of the icons active or selectable at the top and start working with LTspice XVII, first create the new schematic.

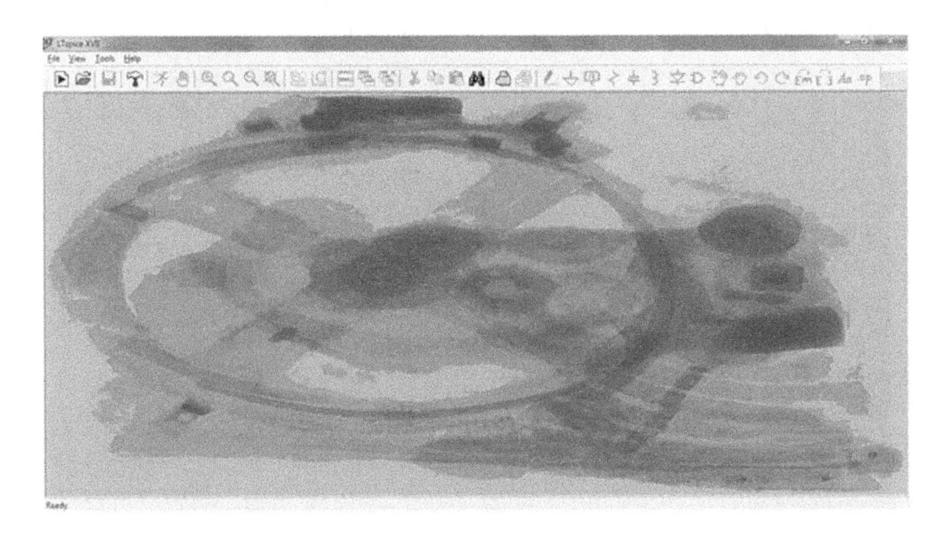

FIGURE 1.2 The Start-Up Window for MS Windows PC.

- **How To Open the Schematic Editor Window** (a blank capturing screen or canvas on which a circuit is drawn using available icons and tools to define the circuit for simulation):

To open the schematic capture, either click on the New Schematic icon or select the option New Schematic from the File schematic menu. This opens the default grey colored fresh schematic window (general-purpose schematic-driven capture/SPICE program) with the expanded menu bar and colored toolbar on the top where the component models are placed and interconnected to define the new circuit schematic along with placing directive statements for simulating the drafted circuit (see Fig. 1.3). The simulator software contains models for several active and passive components/elements/devices which are the parameterized equations for governing their operations (e.g., I-V characteristics of a diode). The bottom-left corner of the schematic window is the Information Bar (message area to provide relevant data information to the user).

A status bar (the Information Bar) can be seen at the bottom where the user can find out relevant information related to measuring electrical quantities, net labeling, dx and dy measurements, etc. By default, the Schematic Editor is having a grey background.

(**Note:** The schematic window's user interface can be modified and adjusted as desired using the **Control Panel** option.)

- **The Schematic Drop-Down Menus and Toolbar Action Icons**

In the new schematic, the schematic menu bar gets extended by left-clicking on the specified menu icon so that more options can be selected from the schematic pull-down menus, for example, the View menu is shown in Fig. 1.3.

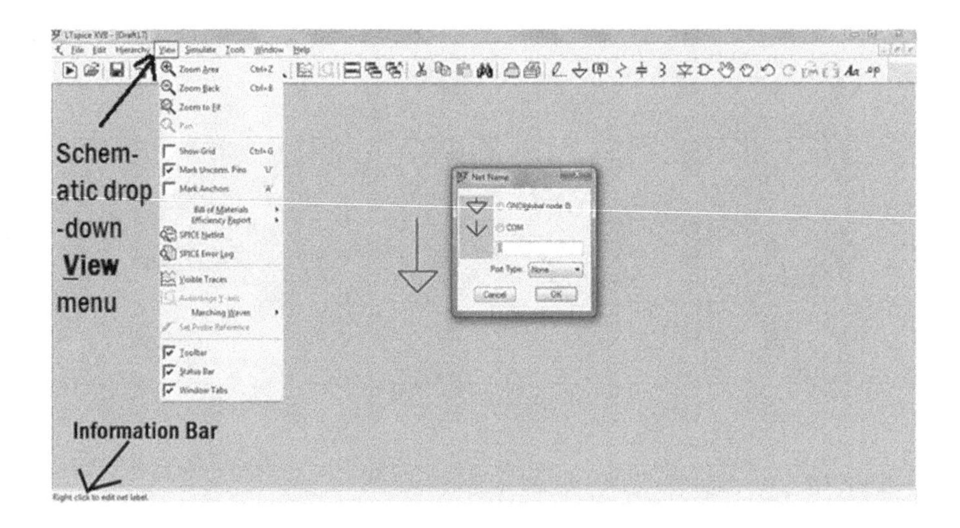

FIGURE 1.3 The Blank Schematic Editor for Drafting Circuits.

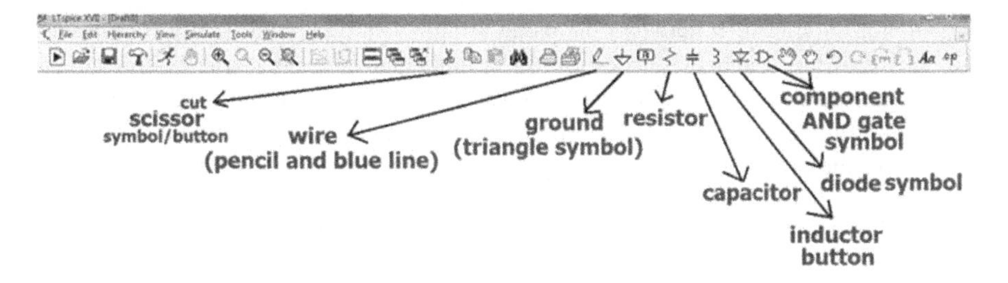

FIGURE 1.4 The Schematic Toolbar Action Icons/Buttons.

Also, all the colored toolbar icons on the schematic interface (see Fig. 1.4) imply that they are in an active state.

The new schematic can be saved as the .asc file or we can start by creating an empty draft. The mouse wheel or the zoom icons on the schematic are used to zoom in or out the drafted circuit.

- **Placing Grid Dots on the Schematic:** Grid dots can be put on the schematic by clicking on the View pull-down schematic menu and selecting the Show Grid entry option by right ticking the check box click or pressing Ctrl + G on a keyboard. Alternatively, insert this option using the Control Panel schematic tool, which can be accessed by clicking the button with a hammer on it or select the **Tools -> Control Panel**. With the Control Panel dialog window opened, first select the **Drafting Options** tab, and then check the box next to the Show schematic grid points field and click the OK to have grid points on every new schematic. The grid points are useful in drafting circuits on the schematic by making components' alignment much easier. The grid facilitates inserting and linking components together on the schematic. Here, we can observe that most of the action icons are now active (colored) and selectable by left-clicking on them.

  (**Note:** If the default grid spacing is too large (schematic components also appear large), click on the Zoom Out icon to reduce the grid spacing.)

- **Changing the Schematic Background Color to White: Refer to Section 3.3.**
- **The Context-Sensitive Menu:** Instead of using the available schematic toolbar, the context-sensitive menu (which appears in response to the user action) can also be utilized with a mouse click (typically by right-clicking) on the start-up window.

1.3 USING A SCHEMATIC EDITOR FOR DRAFTING THE CIRCUIT SCHEMATIC

The schematic editor window is used for circuit design and analysis by entering circuit information graphically which includes placing and connecting components, organizing the circuit schematic, choosing an analysis type, and finally running it. To start drafting a schematic design, all components present in the circuit along with a voltage reference ground are required to be inserted on the schematic using the icon-buttons available at the top of the schematic editor (a blank capturing screen) as shown in Fig. 1.4.

Remember: The schematic circuit diagrams are limited to one page.

1.3.1 Components Settlement

- Placing a resistor, capacitor, diode, and inductor on the schematic (drawing/canvas): First, select a capacitor by left-clicking on the capacitor symbol like button in the toolbar (alternatively, press a letter C on a keyboard). Once selected, as many as desired capacitors can be placed on the schematic (canvas for drawing a circuit) by left-clicking on the schematic.

(**Note:** A spirit image of the component symbol appears on the canvas when an arrow cursor is moved off the symbol into the schematic window. Now, move the component to the desired location (by dragging a mouse without holding any mouse button down) and click a left mouse button to place the component to the schematic. After placing the symbol (with a left-click), the simulator offers to add another copy of the same component type on the schematic when a cursor is moved off the placed component body. This process can be continued until we right-click a mouse or press an Esc key to exit the component placement mode)

Remember: Always rotate (using Ctrl + R) the component symbol before **placing** it on the schematic with a left-click. Also, to exit placing the same component type either right-click or press a different key or press an Esc.

Similarly, push the inductor symbol/button (or press a letter L on a keyboard) and place an inductor on the canvas by left-clicking on the schematic. Click the resistor button (or press R on a keyboard) to place the resistor model. After placing the diode model, the canvas resembles Fig. 1.5.

Thus, it can be stated that to place the components/device models such as resistors, capacitors, inductors, diodes, ground, etc. on the new schematic from the schematic toolbar, first select the corresponding symbol/button by left-clicking, afterward left-click on the empty schematic to place it, and then right-click or press an Esc key to end the function of adding the component and come out of this Add component mode. If a right-click or an Esc is not used, a second copy of the same component appears on the schematic when the user moves a cursor away from (or off) the component body.

- **Components Designator Labels and Values:**
    a. **Component Names:** When the user places components/elements on the new schematic, they all are automatically assigned reference designator

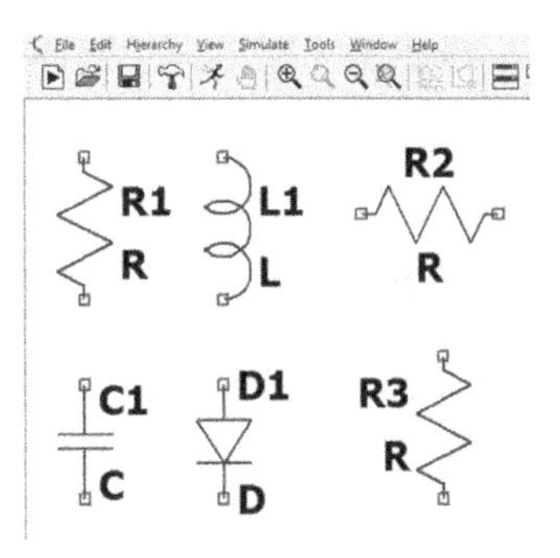

FIGURE 1.5 Placing the Components on the Schematic Window.

labels so that the software can use them as unique identifiers, for example, a capacitor placed above on the schematic is assigned label C1.

b. **Component Attributes:** All the elements, when placed, usually do not have numeral values by default, instead the initial values for all resistors, capacitors, and inductors are the placeholder letters R, C, and L, respectively. For example, a capacitor placed above on the schematic is assigned the initial value equal to C. The numerical values can be assigned to the components by typing their parametric values in the appropriate fields available in the component's respective properties/value editor window (Section 1.5).

- **Editing Component Labels to Designate Meaningful Names**

All LTspice XVII components have names, for example, by default, the resistor component names begin with a letter R and are followed by a number. The default (automatically assigned) label can be easily changed by right-clicking on it and entering a desired component name or text which is more meaningful. It is always advantageous to assign purposeful names to the components placed in the circuit which can have an important effect on the user's understanding of them and assist in writing the simulator directives easily. As stated earlier, the simulator automatically assigns names to the components when they are placed on the schematic. Right-click on the **name** text, say R1 (**not** on the component body or value text) to open the component designator window. Now, the user can assign a different name to the model by changing the default name/component label as shown below in Fig. 1.6.

In the **Enter new reference designator for R1** window, the user can modify the default component label R1 (a letter followed by a number) and assign a more meaningful name whenever feels like. The other components such as capacitors, inductors, voltage, and current sources, etc. are handled in the same way.

(**Note:** No two same component types can be given the same name.)

- **Editing Component Attributes** (The action is needed for LTspice XVII to give simulation results successfully): To modify an object's attributes,

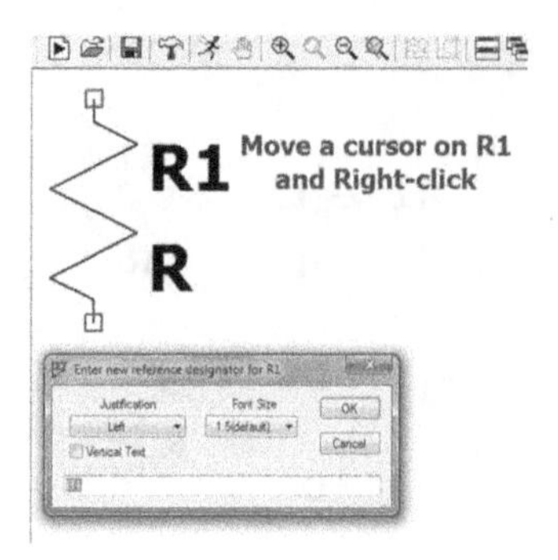

FIGURE 1.6 Changing the Name of the Component.

right-click on the object or the object's initial value (i.e., the C in case of an above capacitor). Depending on the nature of the component, an editor window with a specific dialog box appears on the screen where the user can change the component parameters.

- **Assign Numeric Values to the Components**
  To simulate the schematic circuit diagram successfully, the user must assign numeric values in the Value field of the components either in case-insensitive default units (the available data labels specifying the default units include V, A, Ω (ohm), Hz (hertz), J (joule), W (watt), s (second), deg (degree), etc. so that the user can just enter the numeral 5 to set a value of 5 V (volts) to the voltage source component) or any other specified abbreviated unit entered manually by the user. As LTspice XVII also knows value suffixes or unit prefixes that are not case-sensitive as shown below in Table 1.1, the user can also assign component values with units other than the default by using standard notations (with an exception of using **MEG** or meg to specify 10^6). LTspice XVII always ignores anything that is not a known value suffix or unit prefix.
- **Entering Abbreviated Units:** The user may also write **5V** instead of typing just the number 5 for the voltage source value to abbreviate units.
- **Entering Unit Prefixes (or Value Suffixes):** All numbers in scientific or exponential form can also be entered by using engineering multipliers or value suffix notation so that 1e5 (or 1E5) equivalent to 10^5 (or 100000) can also be written as **100k** or **0.1 Meg**. Thus, to set the resistance value of the resistor model equal to 1000 Ω, the user can enter either 1000, 1e3, or 1k in the component value field. Characters like x10 and ^ are meaningless in the simulator and can lead to errors. To assign unit prefixes, we can use the standard characters like k, m, u, n, etc. as multiplication/scale factors immediately after typing the numeral values (i.e., in LTspice XVII typing **10k** on the resistor model parameter value equals 10 kΩ, **10u** on the capacitor equals 10 μF and **10f** equals 10 femtofarads).

TABLE 1.1
The SPICE Standard Number Notations

Value Suffix (Case-Insensitive)	Numerical Multiplier
T/t	1e12 (equivalent to 10^{12})
G/g	1e9
Meg/meg	1e6
K/k	1e3
mil	25.4e-6
m/M	1e-3
u (LTspice XVII replaces with μ) or μ	1e-6 or 10^{-6}
n	1e-9
p	1e-12
f	1e-15

From Table 1.1, it can be observed that the simulator follows standard notations with an exception of using MEG (or meg) to specify 10^6 because suffixes are not case sensitive (1m and 1M are equivalent to 10^{-3} or 0.001). Also, if a number is written in form 5k3 in LTspice XVII, it implies that the number is equal to 5.3k. This works for any of the multipliers mentioned earlier. It can be turned off by going to the Tools -> Control Panel ->SPICE and unticking the **Accept 3K4 as 3.4K** option.

Remember: Unrecognized letters immediately following a number or value suffix (an engineering multiplier) are ignored in the simulator so that 5, 5V, 5Volts, and 5Hz all represent the same number. Also, M, MA, Msec, and Mohms all represent the same scale factor (.001), and therefore 1M in the value field of the resistor is interpreted as a one milliohm resistor.

(**Note:** Do not use spaces anywhere in the value text to avoid an occurrence of an error during a simulation.)

To conclude:

i. In most cases, a component's attributes can be modified by right-clicking on the component. This opens up its specific editor window as shown in Fig. 1.7 so that the component parameters can be changed. Move a cursor on the component until a pointing finger appears.

Right-click and type in the value of a resistor R1 in the Resistance field box Resistance[Ω]: (the symbol Ω denotes the default unit) under the Resistor Properties option.

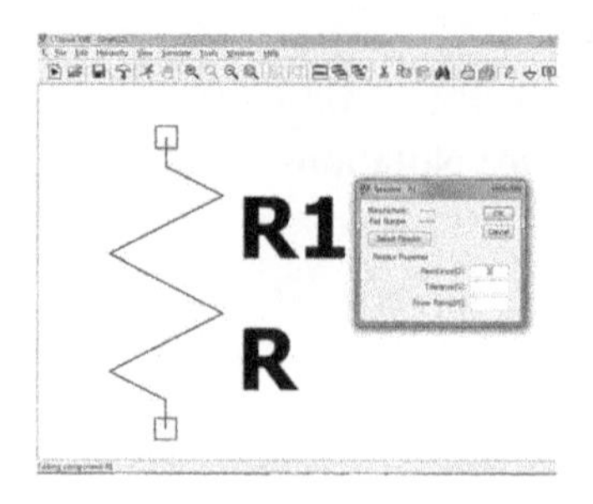

FIGURE 1.7 Setting the Resistance Parameter Value for the Resistor R1.

ii. A numeral value can also be assigned to the components by right-clicking on the value of the component placed on the schematic instead of the component itself. Move a cursor on the value of the resistor R1, that is, on R until a plus cursor changes into a vertical cursor, and then right-click. A different value editor window for a resistor R1 appears on the screen this time as shown in Fig. 1.8.

Replace the value text R by entering a numeric value (number or expression containing operations on numbers while encased in curly braces) for the resistance parameter of the resistor R1 in a value text box at the bottom and click the OK.

To enter 100 Ω, just type 100 in the box as the simulator assumes the resistance in a default unit of an ohm. For assigning value suffixes or unit prefixes other than an ohm, the user can employ a character p for pico, n for nano, u (letter U) for micro, k for kilo, m (letter M) for milli, and MEG for mega. For assigning a 4.7 kΩ (kiloohms) resistance value to the resistor model, use either a conventional American format, that is, 4.7k or a European international format, that is, 4k7.

Similarly, for the voltage source component, it is not needed to put the abbreviated unit V for volts, Hz for hertz, and so on. For a capacitance value of the capacitor model, just enter 1 for a 1-farad capacitor. It is optional to enter 1F but typing the value with the default unit of farad may be incorrectly interpreted as 1 femtofarad by the simulator. Also, do not use spaces anywhere in the value text to avoid any error during a simulation.

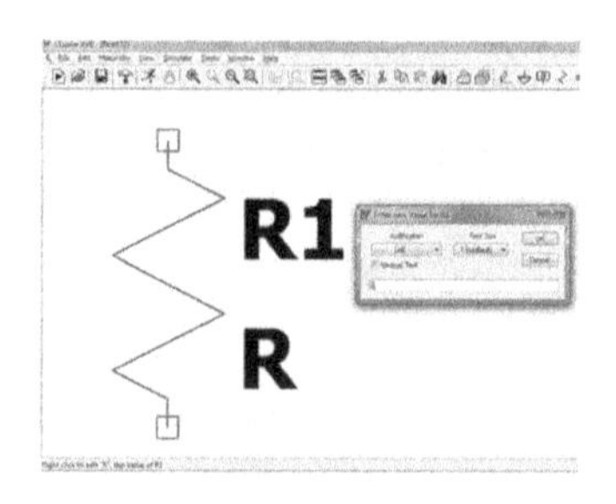

FIGURE 1.8 The Properties Editor Window for the Resistor Model R1.

- **Placing a Voltage Reference (Ground Node)**

The last component that must be present in all the drafted circuits is the voltage reference or common ground node. Left-click on the schematic tool icon that is ground-shaped (or press a G) for placing a ground (the GND or global node 0) at the common point (see Fig. 1.3).

(**Note:** All circuits created in LTspice XVII must contain a voltage reference ground (the GND node) for a simulator to work. SPICE assigns a voltage at this node to be 0, and thereby solves for the other nodal voltages relative to this reference node).

1.3.2 Components Layout and their Passive Sign Convention

The software designates two-end terminals, say terminal 1 and terminal 2, to the default layout of the component models for reference and assumes that a current is flowing into terminal 1 (+ve point) of the device/component, and then out of the other terminal 2 (–ve point). Therefore, a current flowing out of the +ve terminal or entering into the –ve terminal is considered to be negative.

Taking the LTspice components default layout into account, if we rotate a capacitor, resistor, etc. **just once**, their polarity along with the displayed current direction gets reversed as can be seen in Fig. 1.9. This is because all the passive components have two-end terminals and the software represents one-end terminal which is arbitrarily named as 1 to be positive with respect to the other end terminal arbitrarily named as 2. Whenever we place a component, it takes a default position with its end terminal 1 (+ve) on the top and the other end terminal 2 (–ve) on the bottom. Each rotation moves the component clockwise by 90 degrees. To get the end terminal 1 facing left, the user needs to rotate the component (say, resistor here) 3 times from its default position as indicated below to achieve the correct current directions and voltage polarities in the simulation.

a. **Default Orientation/Layout of the Resistor R1 (Vertical or 0 Degrees)**
b. **Horizontal Orientation with Terminals Reversed after First Rotation (+ 90° clockwise)**
c. **After Second Rotation (+ 90 Degrees Clockwise)**

d. **After Third Rotation (+ 90 Degrees Clockwise)** (Fig. 1.9)

As we all know, the universal sign convention in the case of a two-terminal device/component is that a positive current flows into a + plus sign (and out of a –ve sign), and a voltage polarity needs to be consistent with the current direction. LTspice XVII also follows the same sign convention for the passive components and sources, and thus represents a current to be positive if it is entering the component/device from its end terminal 1 and leaving the component/device at its end terminal 2. Also, the software represents a positive voltage at its end terminal 1 with respect to its end terminal 2. For example, if a capacitor is placed on the schematic with its

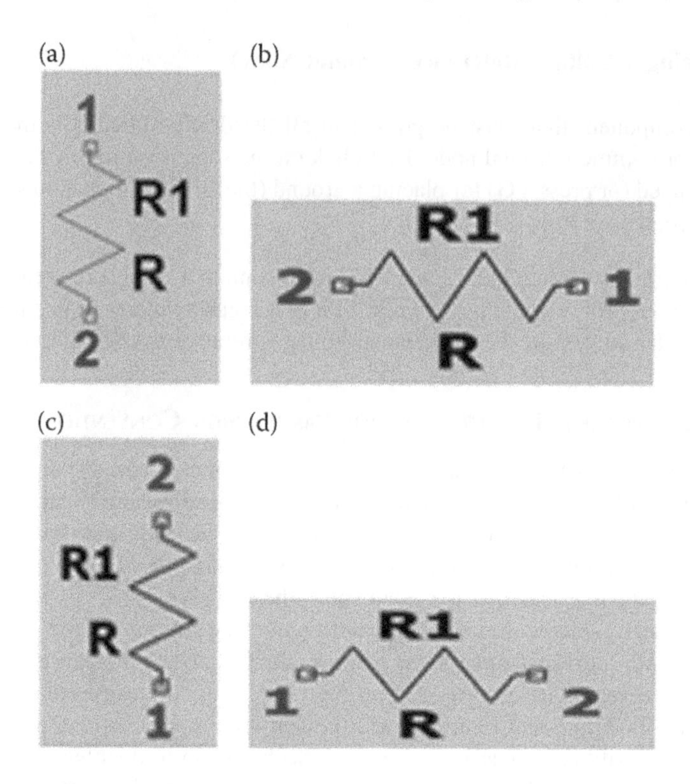

FIGURE 1.9 Change of an Orientation after Successive Rotations.

default layout and its initial voltage is set equal to 10 volts, the simulator puts a 10 V across the capacitor taking the top end terminal 1 to be positive with respect to the bottom end terminal 2.

Thus, to place the resistor component model in a horizontal orientation, rotate it 3 times using Ctrl + R before placing it on the schematic to get the results with accurate polarity.

To conclude, whenever the default layout of the component is changed by rotating it, a new orientation of the components needs to considered and shall be changed accordingly (by rotating them a desired number of times and doing mirror image operations) so that the positive current direction assumed by LTspice XVII matches that of the desired current in the circuit we are simulating.

1.3.3 Components without the Assigned Toolbar Icon and Their Database

- **Selecting the Component Symbols/Models without the Specified Icon on the Toolbar**

The components such as a resistor, capacitor, inductor, diode, and ground can be selected by clicking on their specified/committed icons/buttons in the toolbar, respectively. But all the other components and models, for example, various voltage sources, current sources, BJT models, OP-Amps, MOSFETs, JFET transistors, LEDs, Comparators, IC chips, voltage-controlled voltage source, etc., available in LTspice, contain no separate icon in the Toolbar. Therefore, we need to search them manually in the library of common components. To access these components, left-click on the Component button (AND gate symbol) in the schematic editor toolbar which opens the **Select Component Symbol** dialog window having a wide range of common components and a huge directory of additional circuit elements arranged alphabetically, as shown in Fig. 1.10.

First, find the desired component by scrolling to right on the Select Component Symbol interface, and then to select the component, click on it, and afterward press the OK button (or double click the component). Alternatively, an easy way is to type in the symbol name of the desired component in the Search Bar (text box) indicated on the right, for example, type voltage to get a voltage source (the component of interest). Click the OK once the component appears on the left side of the dialog window to insert the model symbol into the schematic as shown in Fig. 1.11.

The selected component can be moved to any location on the schematic and dropped on it with a left-click. To find the symbol names of various components, click the Help schematic menu and select the Help Topics sub-menu to bring up the following LTspice XVII Help information window as shown in Fig. 1.12.

By default, the **Contents** tab is clicked as can be seen in Fig. 1.13. Click on the expand-button icon to expand all the information as shown below.

Now, select the LTspice® icon and double-click or click on the icon to open up the following sub-menu items (Fig. 1.14):

Now, select the option **Circuit Elements** by double-clicking on it or use the expand-button to open-up the following sub-menu (Fig. 1.15):

From Fig. 1.15, we can observe that a letter (or the symbol name) representing the voltage-dependent voltage source (VCVS) is E. Typing the letter E in the **Select Component Symbol** dialog window gives the VCVS-component model as shown in Fig. 1.16:

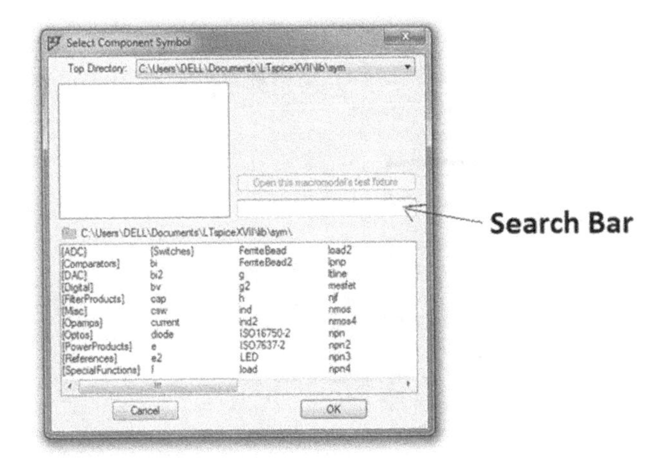

FIGURE 1.10 The Select Component Symbol Dialog Window.

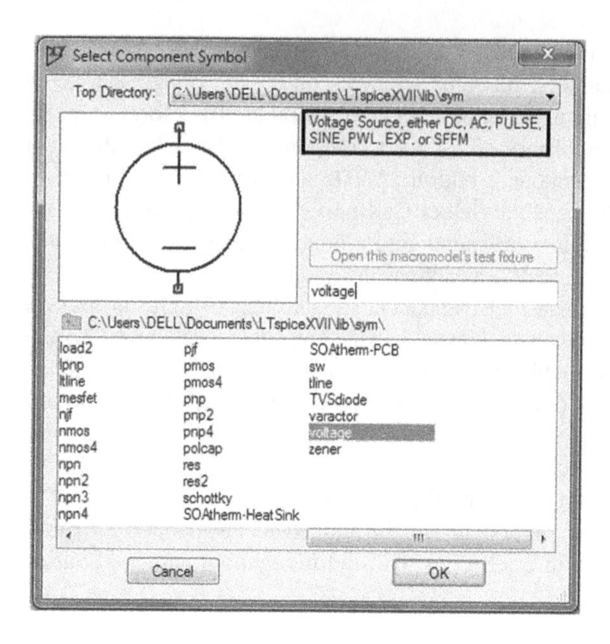

FIGURE 1.11 The Library of Custom Components (Voltage Source Selected).

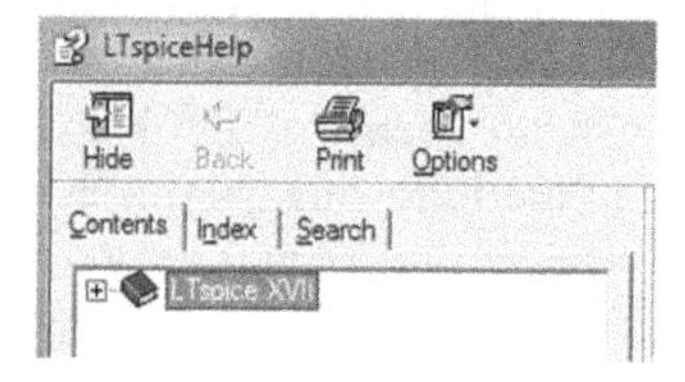

FIGURE 1.12 The LTspice Help Information Window.

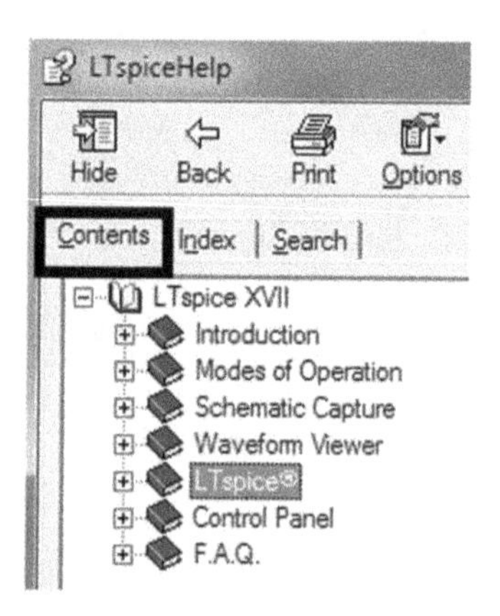

FIGURE 1.13 The Contents Tab.

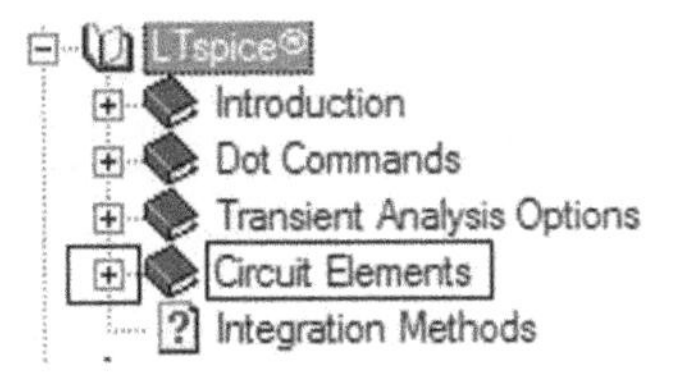

FIGURE 1.14 The Help Expanded Menu.

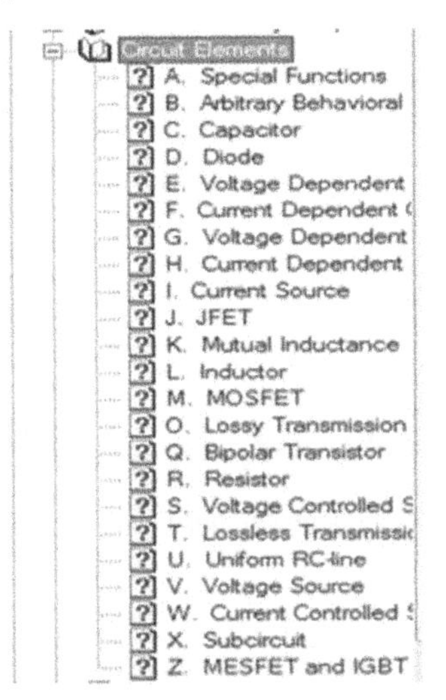

FIGURE 1.15 The Expanded Sub-Menu under the Circuit Elements Category.

- **Additional Components Sub-Directories**
- **How To Access the Battery, Crystal, and Signal Input Sources:** Click the insert AND-shaped component icon (or press F2) on the schematic toolbar. First, make sure that a scroll bar in the **Select Component Symbol** dialog window is to the extreme left and from the additional parts sub-directories on the left-most side (names in square brackets), double-click the [Misc] sub-directory and select the xtal, and afterward hit the OK (or double-click the xtal).

If now, the user revisits the insert component mode, the components of the [Misc] sub-directory are still opened and getting displayed in the **Select Component Symbol** dialog window. To return to the root directory, the user needs to click the yellow-colored **Up One Level** folder icon as shown in Fig. 1.17).

- **Accessing Transformers:** Although there are no separate transformer component models present in LTSpice XVII, the simulator provides the Help schematic menu which contains the required information on how to create it (for accessing the help facility, go to the Help menu, select the Help Topics sub-menu, then select the Search tab and type transformer in the search text box). Alternatively, select the F.A.Q. under the Contents tab to find instructions related to simulating transformers (see Fig. 1.18).

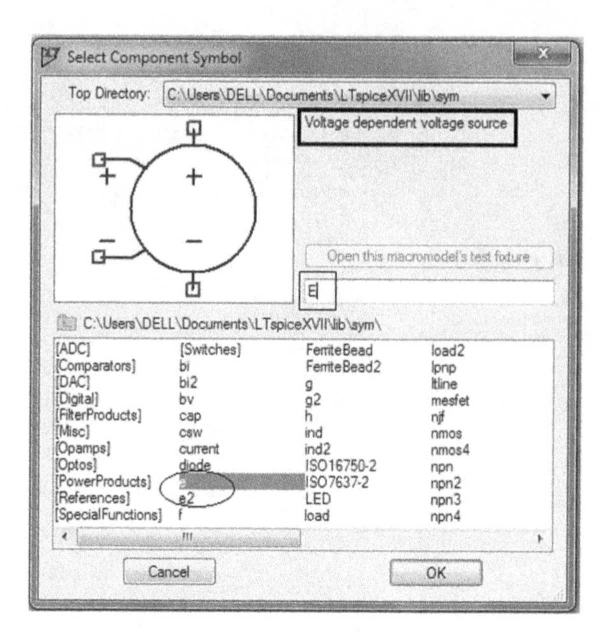

FIGURE 1.16 The Select Component Symbol Dialog Window.

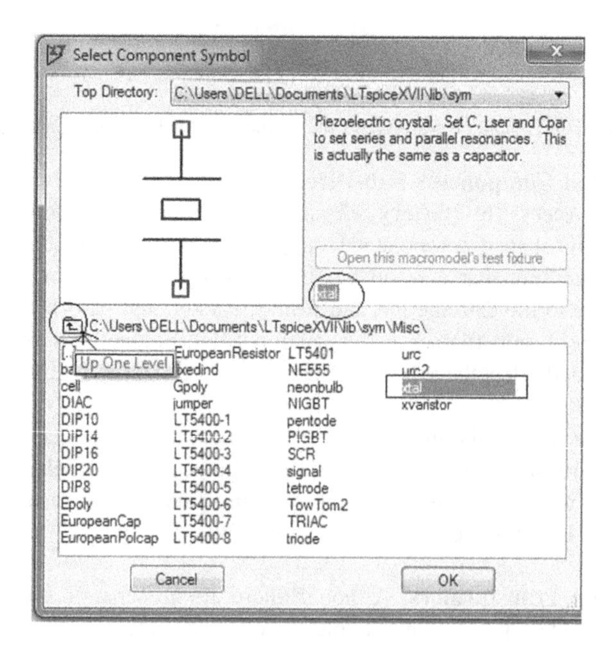

FIGURE 1.17 The Select Component Symbol Dialog Box.

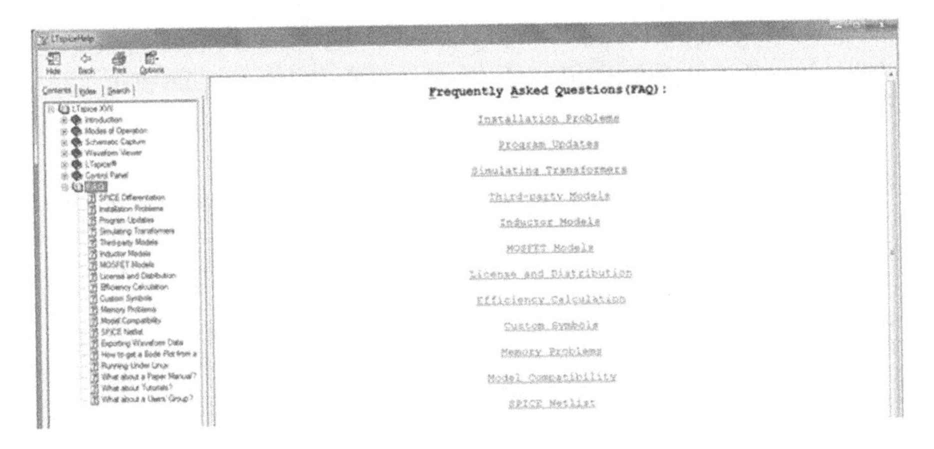

FIGURE 1.18 The Contents Tab under the Help Section.

Hints: First, create two inductors L1 and L2 to be the two windings. Press the SPICE Directive tool (.op icon) to enter the simulator directive and in the box, type in K1 L1 L2 k, where the k is a coefficient of coupling. Normally, use 1 for the k in the case of a toroid or power transformer. Something smaller is used for air-core transformers. After clicking the OK, click a cursor (changed into the command) in the drawing to put the directive command on it (preferably near the transformer). Note that polarity dots were added to the inductors after placing the directive command. A transformer with a 1:3 turns ratio should have a one to nine inductances ratio, that is, proportional to a square of turns ratio because inductances of windings are in the same ratio as an impedance transformation and a square root of inductances ratio is the same as a voltage transformation ratio. The mutual coupling coefficient is set to 1 to model a transformer with no leakage inductance.

Transformer, center-tapped: Create three inductors so that two of them can be connected in series to form a center-tapped winding. The spice command is the K1 L1 L2 L3 1. Here, inductances of the two inductors forming a center-tapped winding are one-fourth of the total winding inductance. For example, a 1:1 (total winding) center-tapped RF transformer should be drafted with inductors of 10 μH, 2.5 μH, and 2.5 μH.

(**Note:** The custom component models can be modified to allow the user to specify additional properties for the components, for example, the generic independent voltage source can be modified to give different waveform shapes other than DC such as sinusoidal waves, pulse waveforms, triangular, and many more arbitrary waveforms)

- **Accessing the Components' Available Database**

LTSpice XVII also provides a database of manufacturer's attributes for some of the components such as diodes, BJTs, MOSFETs, JFETs, independent voltage and current sources, etc. which can be easily added to the schematic diagram. For example, to configure the component diode to one of the manufacturer's attributes, first:

i. Right-click on the diode symbol placed on the schematic.
ii. Left-click on the Pick New Diode option in the diode Properties window.
iii. Left-click on the desired device model name 1N914 as shown in Fig. 1.19.
iv. Left-click on the OK to place the specified diode model 1N914 on the schematic.

Similarly, the available database of transistors can be accessed easily.

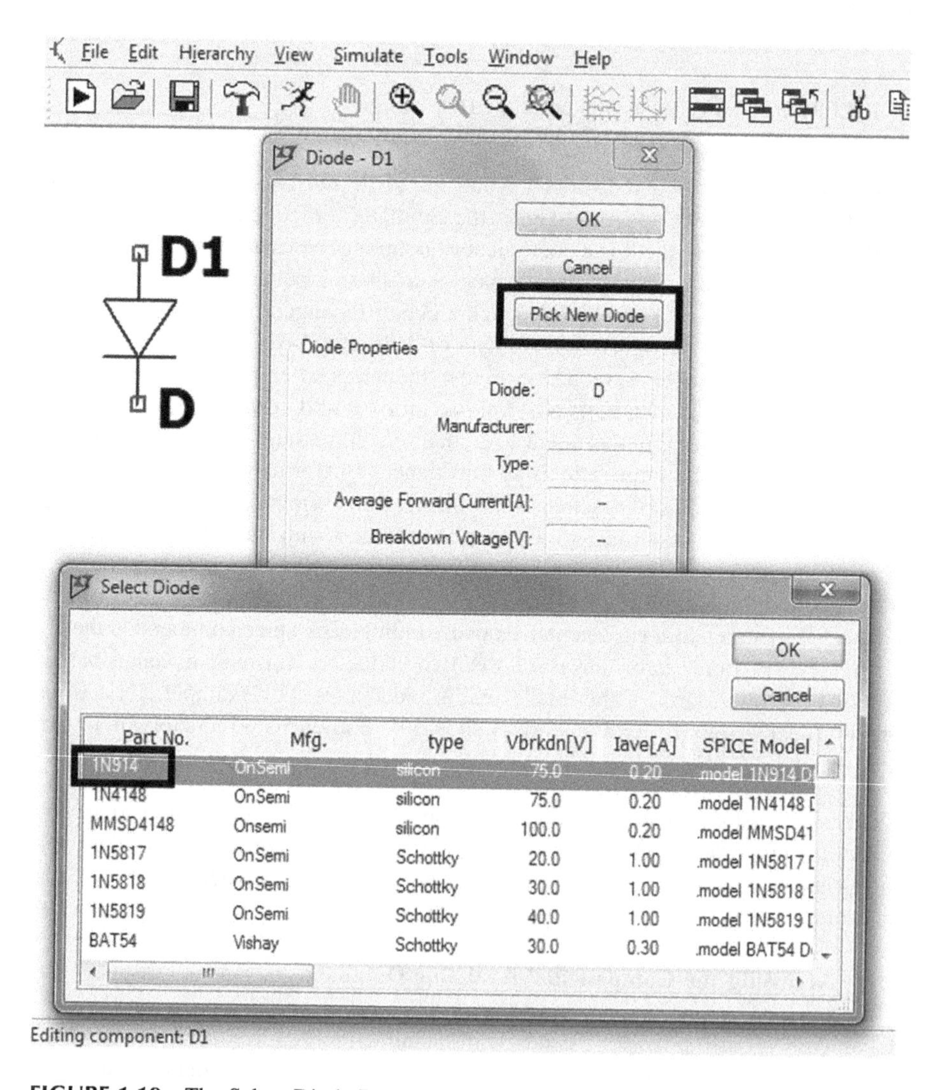

FIGURE 1.19 The Select Diode Database Window.

Thus, the library of models for custom components can be leveraged to quickly build or modify a schematic using realistic components.

The database of a capacitor can be accessed by right-clicking on the component to open up its property editor window and left-clicking the **Select Capacitor** button as shown in Fig. 1.20.

Similarly, a database of the resistor and inductor models can be accessed.

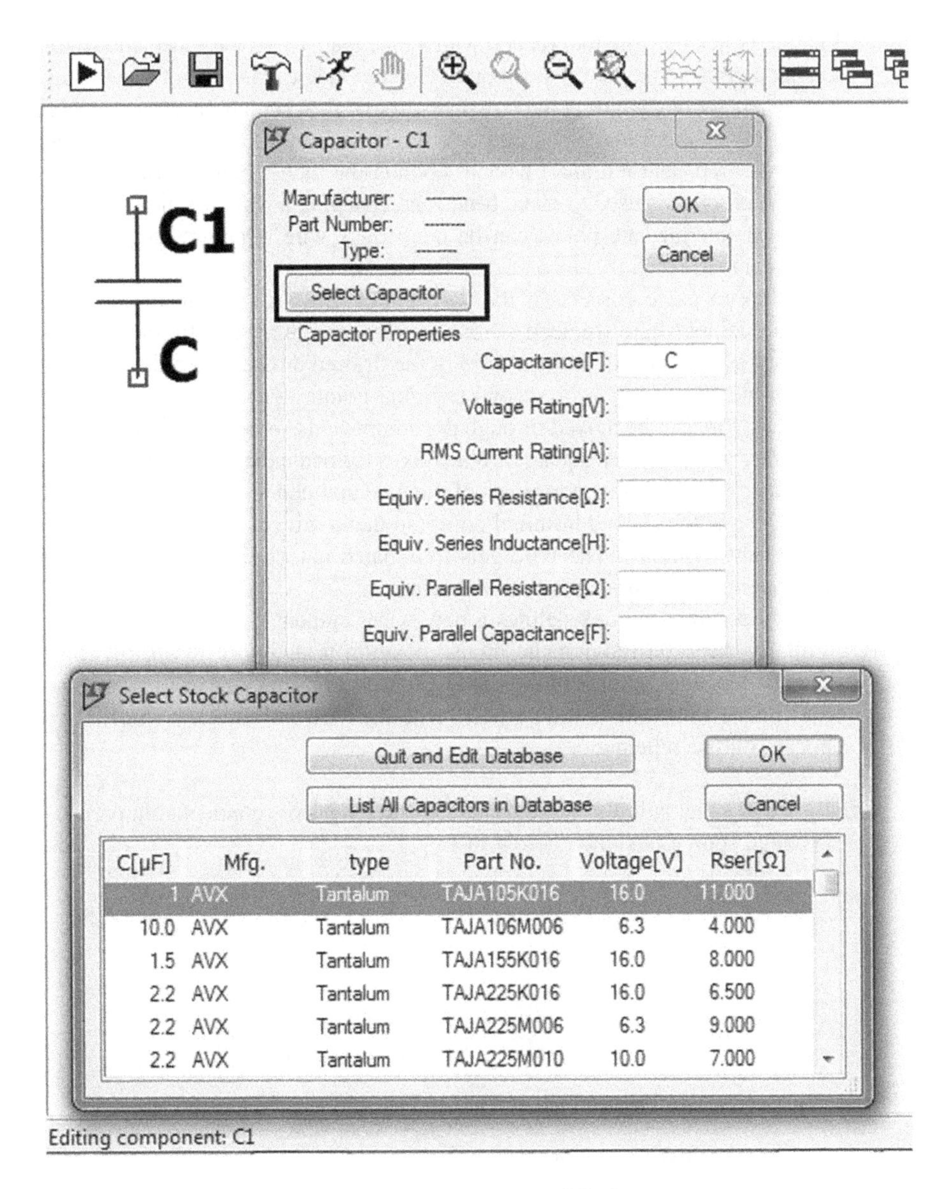

FIGURE 1.20 The Select Stock Capacitor Database Window.

1.3.4 Connecting the Placed Components (Wiring)

The components can be connected by using a key F3 or the add wire button (pencil drawing a line) from the schematic toolbar.

How To Use the Wire Tool To Connect the Components on the New Schematic:

i. First, left-click the wire tool (pencil with a blue line) on the schematic editor toolbar and move the mouse position to the starting location (one of the starting component terminals) using the crosshairs to find it. After left-clicking to start a wire at the first terminal point, pull or drag the wire in the desired direction (to the next component terminal) without holding the mouse button down.
(**Note:** Left-click is used to make bends and changing a wire direction so that a path over intermediate points can be defined. A wire is progressed over the components to reach the unconnected terminals.)

ii. Left-click to place corners (if the wire is not straight, click to place intermediate points where we need to make a 90-degree turn) and fasten wires to other wires. Afterward, pull the wire in the desired direction (up/down/right/left) to the next location or second terminal point.
(**Note:** A wire can be moved through the components. When we anchor a wire to another wire, a closed square or blue box is formed indicating a connection.)

iii. After reaching one of the terminals of the next component, left-click again on it to fasten at the second terminal point (so that a wire can be snapped at the component terminal). The wire gets terminated automatically such that the component is now in series with the wire.
(**Note:** Each mouse click defines a new wire segment.)

iv. Similarly, keep on moving the mouse position to add wire to all the components' terminal points with a click. We need to right-click or press an Esc to terminate an action of adding wires with the wire tool so that no more wire appears on the schematic.

A partially wired schematic for a circuit having three resistors connected in parallel to a DC voltage source is shown below in Fig. 1.21.

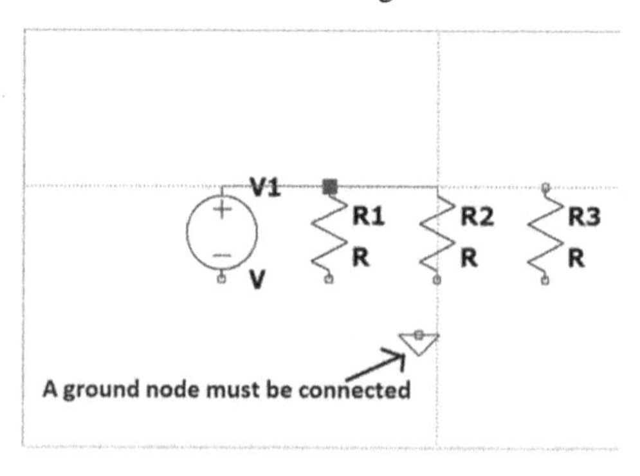

FIGURE 1.21 Partially Wired Parallel Circuit Schematic.

(**Note:** Press a right mouse key or an Esc key to stop wiring or deselect the pencil tool.)

IMPORTANT: For making diagonal connections, the user needs to hold down a **Ctrl** key while drawing lines with a pencil tool. A left mouse click starts a wire segment.

Last but not least, to make a connection to an existing wire, just left-click on it. Also, a blue-colored square block appears at the point where 3 or more wires are interconnected.

1.3.5 Additional Component Editing Features

- **Delete Extras on the Schematic:** If there is any mistake, use the Cut (scissors icon) schematic tool to delete the unwanted parts. First, click a scissors icon on the schematic toolbar, then left-click on the unwanted component to delete it, and afterward press an Esc or right-click to quit deleting. Alternatively, right-click on the schematic (not the component itself) to access the available context-based schematic menu. From the Edit menu, select the Delete sub-menu option to change a cursor into a scissors icon. (Note: Left-click and drag to delete an area or any portion or entire circuit. Press a right mouse key or an Esc key to deselect the Cut schematic tool.)
- **Moving the Components Around on the Schematic:**
    - **How To Move the Components after Placing Them on the Schematic Window:** Select the white hand-shaped Move tool (with spread fingers) from the schematic toolbar, left-click on the component or the component label, and release the mouse button. Now, move the component to the desired location by moving (dragging) a mouse with no buttons pressed. It is required to right-click on an empty area of the schematic to cancel the move function.

A Few Tips:

- The Move tool (a white hand with open fingers) moves components and wire segments individually when the user left-clicks on them. In other words, it can be said that the Move tool moves the component out of connection points or nodes.
- The directive statements, net labels, and text can also be moved using the Move tool.
- The group of components or entire circuit diagram can be moved by first drawing a select box with the Move tool hand. First, position the Move hand tool where you want to start a select box, then click and hold a left-button and drag it (while holding the button) to select the desired components, and afterward release the mouse button. Now, the user can move or re-position the selected components by moving (dragging) a mouse with no buttons pressed.
- The whole circuit can be moved just by moving a cursor over the diagram, left-clicking, and dragging the circuit while holding a left mouse button.
- The Move button can also be accessed from the Edit menu on the schematic menu bar.

- How to drag components after they are connected: Even after placing and connecting the components on the canvas, the user can still change their position by dragging them. Left-click on the Drag icon (a hand symbol that looks like a fist) in the schematic toolbar, and afterward left-click on the component we want to drag and move the component around to re-position it. Right-click on an empty area of the schematic to cancel the drag function.

Important to Note: The drag tool drags wiring along with the component. Dragging leaves wires connected.

- **How To Rotate:** The components can be rotated after selecting but **before** placing them on the schematic by pressing Ctrl + R on a keyboard.

- **How to Mirror Components:** We can mirror components across the vertical by pressing Ctrl + E, while the component is selected (i.e. before placing the component). For example, it can be used to get an NPN transistor with the emitter down and the base to the right.

(**Note:** Rotating and mirroring a component does not affect the circuit behavior, it only allows for clean and readable circuits so that the user can see the connections clearly.)

Points to Remember:

LTspice XVII plots a voltage on the **wire/node** or current through the **component** whenever they are clicked by the user instead of selecting them (**component/wire/node**) for modifications. Therefore, before moving, dragging, or deleting components in LTspice XVII, first select the specific tool, that is, the Move, Drag, and Cut, respectively. Then, the user can select the component/wire/node by clicking on them. Multiple components can also be selected by dragging a box about them.

The program remains in the Move, Drag, or Delete/Cut mode until a right mouse button is clicked or an Esc key is pressed. All schematic edits can be undone or redone.

- **Placing the GND Node:** Place a ground point by clicking on the Ground tool button . It represents the circuit common global node 0.

- **Annotating Schematics:**
  i. **Adding informative comments using a text:**
  Placing a text on the schematic: Place a text on the schematic by using the tool as shown below in Fig. 1.22.
  The aforementioned action only adds some relevant information to the schematic. Therefore, the text placed on the schematic makes no electrical impact on the circuit and its simulation.
  ii. **Adding Graphical Annotations** (Available Under the **Draw** Sub-Menu in the Edit Schematic Menu) **with a Left-Click for Better Illustration of the Circuit:**
  - **Drawing a Line:** Draw a line on the schematic. Such lines have no electrical impact on a circuit, but can be useful for annotating the

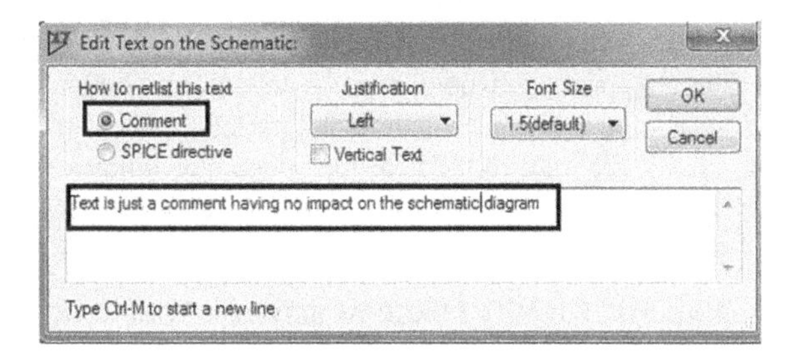

FIGURE 1.22 The Highlighted Comment Field Button.

circuit with notes. For example, to highlight a transformer core, we need to draw two graphical lines.

- **Drawing a Rectangle:** Draw a rectangle on the schematic. This rectangle has no electrical impact on the circuit but can be useful for annotating it with notes.
- **Drawing a Circle:** Draw a circle on the schematic. This circle has no electrical impact on a circuit, but can be useful for annotating the circuit with notes. When inserting a circle into the schematic with a pencil symbol, click a left mouse button and release it. Now, drag the symbol (to rotate it about the origin) by moving a mouse (with no button pressed) to start a circle, size the circle as desired (two anchor points around a circle drawn represent where the element starts and ends), and left-click again to finish the circle. Click a right mouse button to end the selected function of drawing a circle.
  Similarly, all the other symbols are drawn in a way as stated earlier.
- **Drawing an Arc:** Draw an arc on the schematic. This arc has no electrical impact on the circuit but can be useful for annotating the circuit with notes.

(**Note:** Graphical explanations to the schematic by drawing lines, rectangles, circles, and arcs snap by default to the same grid as used for electrical contacts of wires and pins. But, holding down a Ctrl key while positioning these, snaps them to the finest grid available. A right mouse click is used to terminate the annotating functions.)

1.3.6 Adjusting the Schematic Circuit Position

After creating a complete circuit diagram on the schematic, the position (where a whole circuit is drawn initially) of the organized drawing can be changed within the schematic canvas as follows:

i. The **Zoom full extents** schematic tool icon is used to re-center and re-size an entire circuit diagram on the canvas (or capturing screen) so that every placed component, text, simulation commands, and simulator directives are visible to the user.

ii. To move a drafted circuit around on the canvas: Place a plus cursor anywhere on the circuit diagram, press and hold down a left mouse button to grab it, and afterward drag a mouse (while still holding its left button) in any desired direction. The circuit diagram drafted on the schematic follows a mouse cursor. After completing the manipulation, let go of (i.e., release) a left mouse button.

1.3.7 Transferring the Circuit Diagram between Schematics

The user can do the task of transferring (copying and pasting) the component(s) or entire circuit diagram between schematics by using the Duplicate sub-menu in the Edit schematic menu or the Copy icon in the schematic toolbar to invoke a duplicate command.

- To copy and paste the group of components (or full circuit drawing) from one source schematic into another destination schematic:
  - Open the source schematic file, say voltage divider containing the circuit drawing needed to be copied as well as the new destination schematic file, say Draft31 where the copied circuit is desired to be pasted.
  - First, invoke a duplicate command (use an F6 or Ctrl + C or the Copy icon) in the source schematic so that a crosshair mouse pointer changes to a duplicate symbol (two pages together).
  - Left-click to select the component we want to duplicate. For selecting a group of objects or an entire drawing, left-click at an appropriate position (outside the group or drawing), and then drag/draw a box around the group. After releasing a left mouse button, drag the floating copied image to move it in an upward direction to take it out of the active window and get it pasted in another schematic with a left-click as shown in Fig. 1.23.

To paste the copied circuit into a target schematic:

- Once a duplicate image of the object or drawing is copied (Fig. 1.23) and the floating image appears, simply click on the blank already opened .asc destination file icon saved as Draft31 by default (see Fig. 1.23) or New Schematic icon to open the target window, then navigate to a location on the target

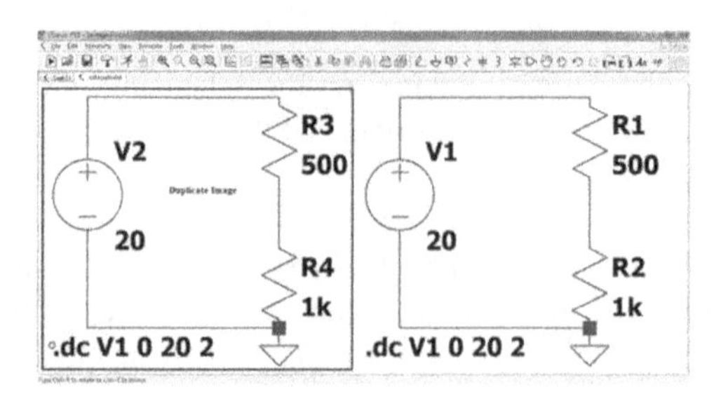

FIGURE 1.23 A Duplicate Image of the Drafted Circuit Diagram.

schematic where the copied circuit is desired to be pasted, and afterward left-click again for placing. In the Windows platform, both schematics must be in the same invocation of LTspice XVII.

1.4 NODES NOMENCLATURE

- **The Default Nodes Nomenclature:** A SPICE program cannot directly understand the schematic diagram or any other form of graphical description because it is a text-based computer program. It can only understand the schematic diagram (graphical description) when a circuit is described in terms of its constituent components and connection points (nodes). Each unique connection point in a circuit is described for SPICE by a node number (N001, N002, etc.) so that the simulator should easily come to know what's connected to what. These numbers can also be considered as wire numbers instead of assuming them as node numbers. A ground node (global circuit common node with reference number 0) with a special synonym, the GND, must be present somewhere in the circuit. All the points electrically common to each other in a circuit constructed for testing and simulation are designated by the same reference number.
  (**Note:** Nodes names may be arbitrary character strings, and therefore 0 and 00 are considered as two distinct nodes.)
- **Displaying the Node Name:** But the reference numbers of the respective nodes assigned by default are not immediately obvious in the circuit and it is very difficult to tell what node has what number. It is only when a cross-hair cursor is brought relatively close to the circuit node, say a generic node N002, the respective node's number (voltage across it also gets displayed but after the simulation) is visible at the bottom left corner of the schematic window in the **Information Bar** as the action displays the text This is node N002. .

  (**Note:** A ground node must always be provided in any schematic for the simulator to work successfully without an error.)

To have a better understanding of the simulation results, it is good to assign appropriate and meaningful names/labels to the various circuit nodes in alphabetic or numeric order as shown below in Fig. 1.24.

The details of how to label various nodes are shown in the following discussion:

1.4.1 INSERTING NET LABELS (THE NODE NAMES)

LTspice XVII automatically labels the circuit nodes or points with specific numbers, but the user can also purposely label them with more significant and logical names to avoid any confusion. Thus, it is always preferable that the user labels anode with a specific name so that an arbitrary one for the node does not get generated by the simulator itself in the netlist files. Labeling the circuit nodes makes it easier for the user to find them in the simulation results. The Label Net schematic tool is used to label the nodes/nets (connection points) on the drafted

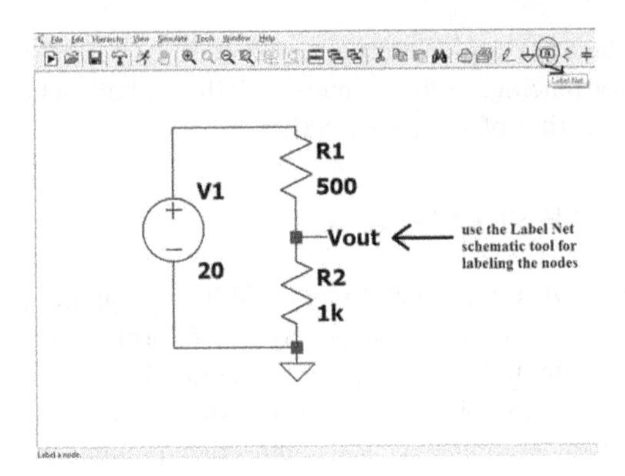

FIGURE 1.24 The Circuit Schematic with the Net Labeling.

circuit schematic so that the user can interpret simulation results quickly. Left-click the Label Net icon to open the Net Name dialog window.

The Net Name dialog window provides three choices for assigning a name/label to the nodes, two radio buttons with predefined graphical symbols (the GND and the COM), and a text box where the user can enter a desired name for the node/net. The graphical symbol GND represents a ground node with the special global net name of 0 and carries a great significance. Whereas, the COM assigns the node with a net name of COM and is of the least importance.

- **How to Place a Net Name/Label on the Specific Node:** In the Net Name field text box, type a meaningful name for an input nodal point, say Vin and left-click the OK as shown below in Fig. 1.25.

The typed name with a small square box (anchor point) **Vin** gets attached to a cursor as shown in Fig. 1.26.

Move the attached anchor point (by dragging a mouse without holding any button) on a wire (or node) between the voltage source V1and a top-end terminal of the R1and left-click to anchor it to the wire (or attach it to the input node) to label an

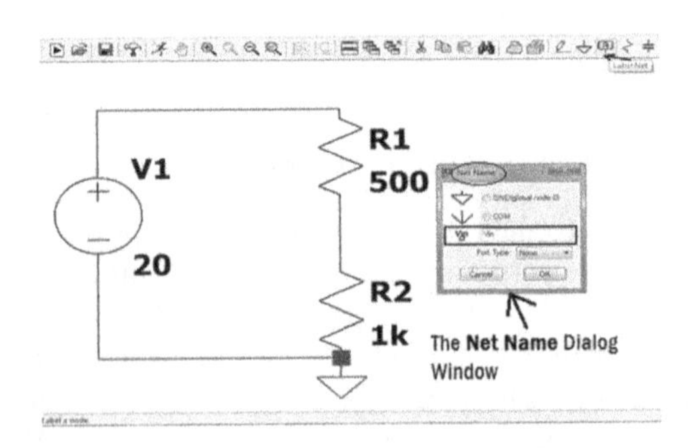

FIGURE 1.25 Labeling the Nodes using the Label Net Schematic Tool.

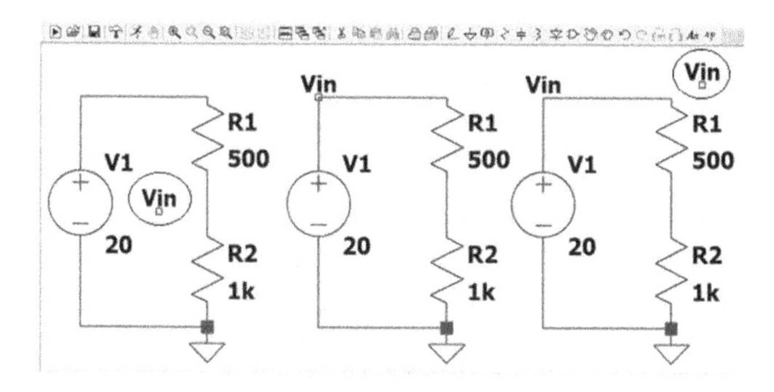

FIGURE 1.26 Labeling an Input Node with the Net Name.

input node between the V1 and R1. Then, right-click to stop the net labeling mode. The net names can also be removed with the Cut/Delete tool.

Also, a small length wire segment can be first drawn horizontally at the junction or connection point between the V1 and the R1, and then the net label Vin is anchored to the drawn extra wire for better visibility and appearance on the schematic.

Last but not least, labeling the nodes or points in the organized schematic circuit proves to be extremely useful for replacing wire connections in complex and very large circuits as well as for identifying them while plotting. Assigning the same name to the two or more nodes shorts the nodes together although no wire is placed.

1.5 CONNECTING NODES TO COMPONENT TERMINALS WITH THE NET LABELING (NO WIRING)

The net labeling feature is required to connect a nodal point to the points (needed to be connected to the same node) in the circuit diagram where wiring makes the circuit appearance untidy, for example, in OP-Amps. The net labeling feature (naming the circuit nodes) makes the complex circuit schematics look quite simple by replacing wiring connections with a linking technique where the same name is assigned to all the circuit nodes or points sharing a common connection. This way, all the required terminals of the devices can be interconnected with the net labeling to avoid drawing unnecessary wires.

1.6 SAVE THE SCHEMATIC FILE AND COPY TO AN MS-WORD DOCUMENT

- **Copy and Paste the Circuit to an MS-Word Document:**
  To copy the circuit diagram drafted on the schematic to any document, do as follows:

  On the schematic menu, left-click the **Tools -> Copy bitmap to Clipboard** options. Close the output text or plot window displaying the simulation results, if opened.

- **Save the Schematic File:**
  - **Saving with the Default Name:**
    Choosing the **File -> Save** option, the user can save a .asc file with the default name (Draft1, Draft2, and so on) assigned automatically at the default location by LTspice XVII.
  - **Assigning a User-Defined Name:**
    Use the **Save as** to change the name and location of the schematic file containing a circuit drafted in LTspice XVII. The user should save the circuit schematic by first left-clicking the **File** (schematic menu) -> **Save As** (sub-menu). Navigate to the desired folder or create a new one and enter a filename, then afterward click the **Save** option.

To conclude: On the schematic menu bar, go to the **File -> Save As** to begin saving the drafted schematic file with an extension .asc and type Schematics. Select an appropriate name and location for saving the file and click the **Save** button.

The saved file can be opened later by clicking on the **File -> Open** option and locating the file by scrolling down or typing its name. Therefore, the user should save the drafted schematic files regularly and must use different filenames for saving them.

2 Simulation Types and Waveform Viewer

2.1 VARIOUS SIMULATION ANALYSIS TYPES

After completing a circuit topology by drafting it schematically, the type of analysis needed to be performed must be defined in order to run a simulation.

2.1.1 Analysis Types

LTspice XVII runs simulation by performing several types of analysis, which include DC operating point, transient analysis, AC analysis (small-signal frequency response analysis), DC sweep, noise, temperature, DC transfer function, parametric, etc.

When an analysis is specified by the user to run a simulation, depending on the analysis type, a text command starting with a period or dot (called a dot command) replaces a cursor pointer which is further needed to be placed on the schematic.

LTspice XVII rich collection of analysis types and their applications are as follows (Table 2.1):

A brief description of SPICE analysis methods for simulating the drafted circuit is as follows:

- **The DC op pnt (the .op)**

A DC operating point analysis is used to check biasing and DC levels (as displayed by a DC multimeter) in electronic circuits. It is the most basic analysis which solves the purpose of computing DC operating point voltages and currents at each node and through a branch (component) of the circuit, respectively. It performs a DC operating point solution (with capacitances open-circuited and inductances short-circuited) as a part of another analysis to find an operating point of the circuit.

The results of the analysis appear as a list in the **Operating Point** dialog box that pops up after a simulation is run. After closing the .op simulation results window, when we point a cursor on the component, a current flowing through it appears on the schematic Information Bar. On the other hand, pointing a cursor on the circuit node/point displays an operating point DC voltage on the Information Bar. In the analysis, capacitors as treated as open circuits, inductors as short circuits, and all the AC sources are considered disconnected.

(**Note:** In the analysis, there is no need to enter any argument.)

TABLE 2.1
SPICE Analysis Types and Their Foremost Applications

DC opn pnt	Determines the DC conditions or static characteristics of devices such as biased transistors, diodes (i.e., operating region values)
	Generates a list of DC node voltages and branch/loop currents of an electric network
Transient	Plots the time response characteristics of any circuit (as like an Oscilloscope)
	Describes the behavior of a circuit to an AC input
DC sweep	A transfer function of an amplifier, I-V curve of diodes, DC characteristic curves of transistors and other devices
AC analysis	Frequency response (gain magnitude and phase in dB versus frequency) of a passive or active filter
	It provides the frequency response of a circuit as a Bode plot so that an output waveform displays an amplitude and phase across a specified frequency range, plots frequency response on a Cartesian coordinate plane with a real and imaginary axis, or generates a Nyquist plot
	Effect of changing the capacitance or any other parameter at a single frequency in combination with the .step
	Impedance as a function of frequency
	A bandwidth of an amplifier, or noise considerations
	Computes output variables such as voltage differences between specified nodes (or one specified node and ground), current output for an independent voltage source, element branch current, impedance, admittance, hybrid, scattering parameters, and input and output impedance/ admittance as a function of frequency
Noise	Examines the response of a circuit under noise conditions to assist in circuit analysis It is also possible to integrate noise over a selected bandwidth
	Performs a frequency-domain noise analysis (shot, thermal, and flicker (1/f) noise) together with an AC analysis statement
	Plots a noise voltage density (in units of volts per square root hertz) over a 1 Hz bandwidth for an output/input noise or any noisy component like a resistor, diode, or transistor. It is also possible to integrate noise over a selected bandwidth
DC transfer	Input and output resistance of an electric network
	Input impedance (input voltage/input current) and output impedance (output voltage/ output current) of an amplifier
	Transfer function or low-frequency gain: output/input conversion ratio (ratio of an output variable to an input variable)
	Output node voltage (or Thevenin's voltage at the open-circuit terminal)of a network for a given (determined) input value
	Output resistance (Thevenin's small-signal equivalent resistance) of a network

- **Transient**

A transient analysis computes transient output variables as a function of time over a user-specified time interval (initial conditions are automatically determined by a dc analysis or can also be specified by the user and time-independent sources are given dc values). A non-linear time-domain analysis is performed so that results are displayed in such a way that an independent axis (x-axis) is timed in seconds and a dependent axis (y-axis) can display any electrical variable (or its mathematical expression).

To conclude, the analysis numerically solves differential equations describing a circuit and uses the simulator as an oscilloscope to observe output waveforms in a time domain. The analysis assists the user in finding a distortion in an output wave, running spectrum (FFT) analysis, viewing actual impedances, and powers delivered/dissipated. A transient analysis shows results in a way an oscilloscope displays waveforms.

- **AC Analysis**

A small-signal (linear) AC analysis computes AC complex node voltages as a function of frequency using an independent voltage/current source as a driving signal. Initially, a DC operating point of the circuit is determined, and then a linearized small-signal model for all the nonlinear devices present in the circuit is established using the computed operating point. Finally, the generated linearized circuit model is simulated for a specified range of frequencies.

The **(none)** function radio button tab option is selected to set an AC version of the independent voltage/current source components (to be used as an input)and amplitude and phase values are specified in the AC Amplitude and AC Phase fields text boxes, respectively. This is because, the source's type (such as sine, pulse, exponential, etc.) does not affect an AC analysis. Further, in an AC analysis and simulation, the user needs to perform several settings such as an x-axis scaling and specifying a type of sweep (linear, octave, or decade), providing simulation points and a frequency range. The analysis results are plotted in the waveform viewer window as magnitude and phase responses over frequency to observe Bode plots (by default) for circuits including amplifiers, attenuators, filters (active or passive), etc. By default, the response is in dB relative to 1 volt on the source. Thus, the analysis computes a small-signal frequency response of a circuit.

There is one more sweeping option used for plotting complex data as a function of a stepped parameter, the List, which allows the analysis at a single frequency (the syntax is .ac list <Freq>) when run in combination with the .step directive statement.

- **DC Transfer (.tf)**

A DC transfer analysis performs a small-signal DC transfer function computation of the circuit's node voltages or branch/loop currents owing to small variations of an independent source. The results include a list of output to input voltage/current conversion ratio and input and output resistance/impedance.

- **DC Sweep**

A DC sweep analysis sweeps an input DC voltage/current source of the circuits in a pre-determined range to plot the response of an output variable as a function of the varying input DC source. The analysis is used for plotting and analyzing characteristic curves of devices and models like diodes, transistors, OP-Amps, etc., and permits an evaluation of a DC transfer function.

- **User-Defined Parameters (.param)**

The .param SPICE directive allows the creation of user-defined variables for storing parameter values and useful for associating a variable name with the model parameter value. To invoke a user-defined parameter substitution and an expression evaluation, it is very important to enclose a parameter value (or an expression) containing a variable name in curly braces to tell the simulator that the component parameter value is a global parameter.

- **Repeated Analysis (.step) with Parameter Sweeps**

The .step directive repeats an analysis several times for several values of a particular parameter such as temperature, the model parameter (such as a resistive value or a transistor attribute), the global parameter, the independent source component, etc. This is done to observe how the circuit's behavior changes as the parameter value changes. In essence, the directive automates the process of incrementing the specific parameter in steps and running analysis over the defined range (start value to end value). The variable steps chosen for the parameter to be swept may be linear, logarithmic or specified as a list of values.

(**Note:** A DC sweep analysis can be called the .step directive in disguise in which an independent source voltage is incremented.)

- **Parametric Sweep Analysis**

It allows the user to sweep a parameter of the component model (or source) through a range of values while running a transient, AC, or DC sweep analysis. A step with parametric analysis is a multirun process where first the main analysis is set, and then a series of values is specified for the component parameter to be swept. When the main analysis is run, LTspice XVII sets a first value of the parametric variable and performs a simulation, and then after finishing it a next value is set automatically on the circuit and the simulation is run again. This process is repeated until a list of values for the component that are being selected is completed and the results are plotted.

- **Noise**

It performs a frequency domain analysis to compute Johnson, Shot, and Flicker noise types (a noise spectral density per unit square root bandwidth). Noise Analysis calculates the noise contribution from each resistor and semiconductor device at the specified output node and propagates it to the output of the circuit sweeping through the frequency range specified.

- **Temperature Analysis**

A temperature analysis is a parametric analysis that uses the .temp command. The analysis is used to verify the operation of a circuit by simulating it at different temperatures. The effect is the same as simulating the circuit several times, once for each temperature. The temperature values can be controlled by choosing start, stop, and increment values.

- **Fourier Analysis**

The analysis is used to analyze complex periodic waveforms. It permits any non-sinusoidal periodic function to be resolved into sine or cosine waves (possibly an infinite number) and a DC component. This permits further analysis and allows determining the effect of combining a waveform with other signals.

2.2 SPICE SIMULATOR DIRECTIVES (THE DOT COMMANDS)

The user can also add SPICE netlist as a text to the schematic capture for setting simulation options. The hand-written directives can be included to mix the schematic capture with a SPICE netlist by selecting the **SPICE Directive** tool button (the .op icon) from the schematic toolbar. With the **SPICE Directive** tool, the user can set additional simulation options (such as measuring quantities, initializing voltages/currents on a device, doing temperature sweeps, and many more) using valid SPICE command syntaxes, include files that contain models, define new models, etc.

To conclude, LTspice XVII extracts the circuit netlist using two main entities:

i. One, by doing the schematic entry (drawing the circuit diagram graphically) which gives component description to the simulator and informs about what components are created in a real instance, the names they accept, where are they connected, and the values of their parameters in one line. For example, if we place a capacitor on the schematic, the user can search for its netlist by selecting the View -> SPICE Netlist from the schematic menu bar.

   The corresponding line on the netlist looks like the following:

   C1 NC_01 NC_02,

   where the C1 indicates that the component is a capacitor, the NC_01 indicates that the capacitor C1 has two end terminals, the NC_01 and NC_02, and C is the default value (non-numeral) of the capacitor C1. The user only gets to know which terminal is the NC_01 and which one represents the NC_02 by moving a cursor near to the respective terminals.

ii. In another example, place the resistor R1 and capacitor C1 in parallel on the schematic and connect the combination across the voltage source V1. Go to the View -> SPICE Netlist to generate the netlist as shown in Fig. 2.1.

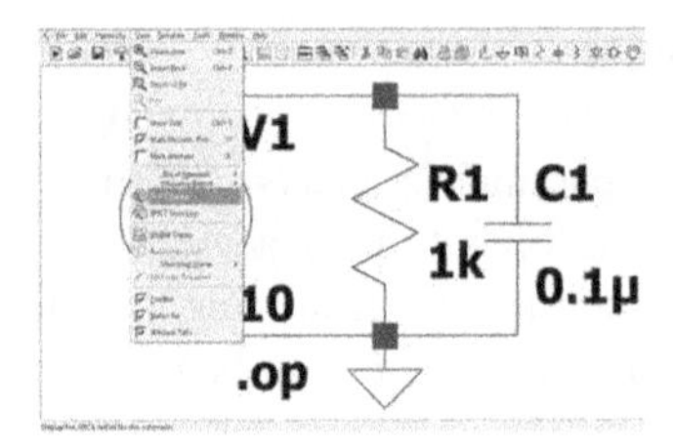

FIGURE 2.1 Generating a SPICE Netlist for the Drafted Circuit.

The text netlist defining the schematic circuit diagram given in Fig. 2.1 is shown in Fig. 2.2.

The externally generated netlist comprising of a list of the circuit elements and the nodes, model definitions, and other SPICE commands, can be edited and simulated. When the schematic is simulated, the netlist information is extracted from the schematic graphical information to a file with the same name as the schematic but with a file extension of .net.

The extracted lines on the netlist files are as follows (Fig. 2.2):

R1 N001 0 1k

V1 N001 0 10

C1 N001 0 0.1u,

where the N001 indicates that the capacitor C1 (or the resistor R1) is connected from the node N001 to the reference node (the GND or global node 0).

The generated netlist can also be edited as follows (Fig. 2.3):

The other mode is to write simulator directives (or dot commands) by using the .op schematic toolbar icon to control the behavior of the simulator at run-time. Thus, it is possible to tell the simulator manually how to run a specified type of analysis along with setting up the parameters through the directive statement. For example, the following directive sets up an AC analysis:

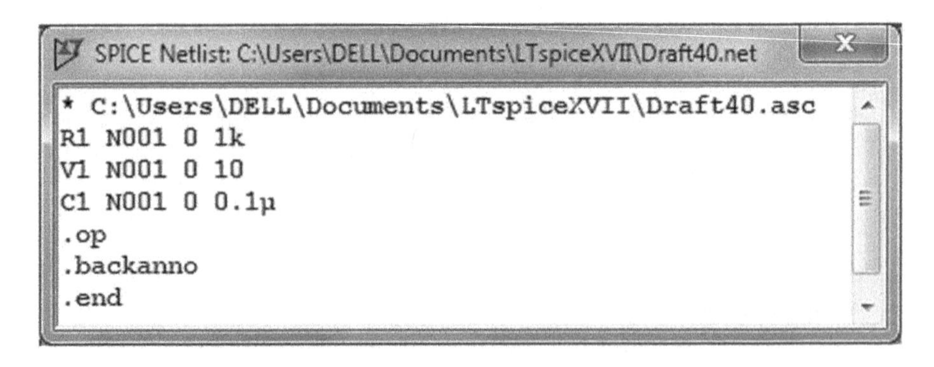
```
SPICE Netlist: C:\Users\DELL\Documents\LTspiceXVII\Draft40.net
* C:\Users\DELL\Documents\LTspiceXVII\Draft40.asc
R1 N001 0 1k
V1 N001 0 10
C1 N001 0 0.1μ
.op
.backanno
.end
```

FIGURE 2.2 The Generated Netlist.

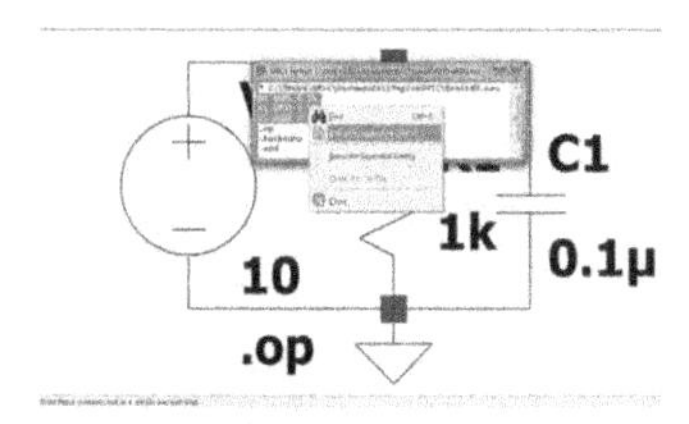

FIGURE 2.3 Editing the Generated Netlist.

.ac dec 10 1 1e3

where the gain is computed between 1 Hz and 1 kHz at 10 points per decade.

Beyond selecting an analysis type, directives are used for many other desirable behaviors such as specifying initial conditions (e.g., charge of a capacitor), analyzing the results of a simulation run, stepping a parameter or customizing a component's model, etc. For example, the .ic directive is used to initialize node voltages and branch currents before a transient analysis is run.

As an example of the .ic directive, consider discharging a capacitor through a resistor in parallel with the capacitor. To examine the rate of discharge of a capacitor over a period of 5 ms with a transient analysis, the directives that need to be written are .trans 5e-3 and .ic V(x) = 1, where the *x* is a name given to the node where the capacitor and resistor models are connected together with reference to the ground node and the required directive to specify an initial charge on the capacitor is simply .ic *V(x)* = 1.

To enter the .ic directive, click on the .op tool button. The following Edit Text on the Schematic dialog box appears (Fig. 2.4) with the SPICE directive button highlighted with blue color so that whatever is written in a text box by the user is considered as the directive.

After typing the syntax for using the .ic directive, press the OK and place the directive statement anywhere on the canvas.

Thus, it can be concluded that the SPICE directive is merely an instruction to the simulation program itself to define what to do with the drafted circuit and is not a part of the circuit.

2.2.1 Summary of Using the Dot Commands

The LTspice XVII dot commands are as follows (Table 2.2):

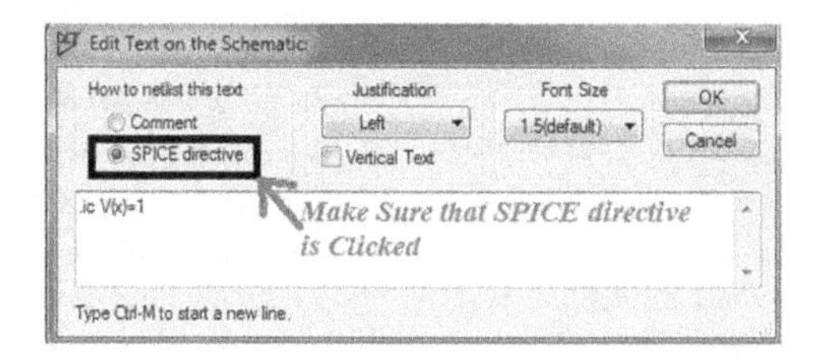

FIGURE 2.4 Mixing the Simulator Directives with the Schematic Entry.

TABLE 2.2
The Dot Commands' Syntaxes

Type	Commands	Comments
Transient	.tran 0 40n 10n	Sweeps a time till 40 nsec and starts collecting data from 10 nsec (nanoseconds)
DC Sweep	.dc V1 4 6 0.2	Sweeps the voltage source V1 from 4 V to 6 V with an increment = 0.2 V
AC Analysis	.ac dec 250 100 10Meg	Sweeps a frequency from 100 Hz to 10 MHz taking 250 points over a decade (20 dB/dec)
	.ac lin 300 10 10k	Sweeps a frequency linearly from 10 Hz to 10 kHz taking 300 points
Temperature Sweep	.step temp 0 100 10	Sweeps the variable/keyword temp (understood as a temperature by the simulator) from 0°C to 100°C with an increment of 10°C
	.step temp list –20 0 25	Sweeps a temperature through the listed values –20°C, 0°C, and 25°C
	.temp 10 20 30 40	Sweeps a temperature for the specified temperatures 10°C, 20°C, 30°C, and 40°C The default temperature for simulating circuit schematics that do not specify a temperature is 27°C
Variable Steps: For Repetition	.step oct V1 2 20 5	Steps the independent voltage source V1 from 2 to 20 logarithmically with 5 points per octave and repeats the main analysis for these values
Parameter Sweep	.param w = 1000 A = 4 .param C = 1	Allows the user to create and store values in user-defined variables that are used globally in place of the component parameter value, provided the variables are enclosed between curly braces, e.g., {w}, {A}, and {C}.
Step with Parameter Sweep	.step param R 1k 2k 100	Sweeps the resistor model's parametric value R (a variable or non-numeric user-defined value attribute enclosed in curly braces) from 1 kΩ to 2 kΩ in incremental steps of 100 Ω Type the value of the Resistor component as {R} in the editor window, i.e., a variable R enclosed in curly braces
Variable Steps: A list of Values	.step param R List 5 10 15	Performs a simulation 3 times with the parameter value of R being equal to 5, 10, and 15
Measuring Values	.meas AC V FIND V(n001) AT 1e6	Finds a voltage V at the node N001 at a frequency of 1 MHz with an AC response
	.meas AC Itotal FIND I(R1) WHEN freq = 100	Finds a current total through the R1 when a frequency value is 100 Hz with an AC response
	.meas TRAN Vo FIND V (Vout) WHEN time = 15 m	Finds an output voltage Vo when a time value is 15 ms with a transient response
Initial Condition Command	.ic V(n002) = 0	To set a voltage on the node N002 equal to zero

LTspice also includes advanced simulation features such as steady-state detection, turn-on transient response, step response, efficiency/power computations, etc. LTspice XVII allows setting a temperature for the simulation in two ways, global temperature for all the devices and local temperature for the selected devices. A local temperature is set by assigning a temperature (constant or parametric value) along with the component's numeric value using the keyword temp. For example, temperature of the resistor model having a resistance of 200 ohms can be set to 100°C by entering 200 temp = 100 in the model's value editor window. Also, a temperature can be fixed for the other components, provided the component's parameters are temperature dependent.

The user can access all the information by using the LTspice XVII help facility. Go to the Help -> Help Topics in the schematic menu bar and type analysis in the Search text box (click on the Search tab) or click on the **Index** tab.

2.3 SETTING AN ANALYSIS FOR RUNNING A SIMULATION

The **Simulate** Schematic Menu **(for choosing an analysis to run a simulation):** To set up an analysis for a simulation, go to the Simulate schematic menu Simulate and choose the Edit Simulation Cmd sub-menu Edit Simulation Cmd.

After setting up an analysis for a simulation, always click the OK. The simulation text command (SPICE directive) gets attached to a cursor which must be placed by clicking somewhere on the drawing to make the selected simulation analysis type effective for the next Run command.

Remember: If we are running a simulation for the first time, clicking on the Run button opens-up the Edit Simulation Cmd dialog window, which allows the user to select the desired simulation type. But if it is desired to change a simulation command after the first run, we need to first select the Simulate -> Edit Simulation Cmd option from the schematic menu. Alternatively, we need to first delete (using the Cut scissors-shaped tool) the SPICE directive text command inserted somewhere into the circuit schematic by the user.

2.4 MANIPULATING THE WAVEFORM VIEWER WINDOW

All graphical traces are displayed in the LTspice XVII waveform viewer built-in window. Except for a DC operating/bias point analysis, all the others output desired information graphically so that the user can have an improved understanding of the results. Thus, it is important to know about the waveform viewer for probing and plotting in the plot pane and edit traces.

2.4.1 Probing and Plotting Electrical Quantities

The different ways of selecting data to be plotted and edited in the waveform viewer after running an analysis are as follows:

(A) Plotting a Data by Probing (Selecting a Data To Be Traced) Directly from the Schematics

i. **For bias point voltage measurements as well as probing voltages for graphical plotting:** After running the desired simulation (close the .op result window in case of bias measurement, whereas a blank waveform viewer should appear on the schematic in case of plotting) do as follows:

   Plotting a voltage: To measure and plot a voltage on the node with reference to the ground, move a mouse appropriately over the respective node (a mouse cursor changes into the red voltage probe cursor) to be measured and left-click.

   (**Note:** LTspice XVII includes an integrated waveform viewer that allows the user to have complete control over the manner a simulation data is plotted by using a data-trace selection method.)

ii. **For probing and plotting current through a device or component:** After probing a voltage successfully, now do as follows:

   Probing a current: To plot a current flowing through a device (wire or node), move a cursor appropriately over the component through which the current is to be measured and left-click when a mouse cursor changes into the current probe cursor (an icon that looks like a clamp-on ammeter).

   Alternatively, move a cursor over a wire while holding down an Alt key, and once the current probe sensor appears, left click to plot a current in the wire.

iii. **To probe and plot a current into a pin of a device having more than two connection points:** First, place a cursor over the pin of interest, and left-click when the current probe image appears.

iv. **Differential voltage plotting:** After running a simulation type involving graphical analysis, a blank waveform viewer appears on the schematic. For plotting a voltage across the device (i.e., between the two terminals), move a mouse on the node representing a positive terminal of the device (across which a voltage is to be measured). After getting the red colored voltage probe symbol on the first (positive) node, left-click a mouse, and then drag the probe to the second (negative) node while holding a mouse button. Now, release the mouse button when obtains the negative black colored voltage probe cursor on the second node (to allow LTspice XVII to differentially plot voltages). A differential voltage gets displayed.

Deleting Waveforms: If the user wants to remove a trace or waveform from the plot pane, press an F5 key (a cursor changes into an image of scissors) or use the Cut schematic tool and delete the appropriate trace label/name at the top of the plot pane (or waveform viewer). Thus, the user can delete individual traces by clicking on the trace's label after selecting the delete command. Alternatively, right-click on the label of a trace (to be deleted) and select the option **Delete this Trace** from the waveform viewer menu. Right-click or press Esc to end the Delete function.

(**Note:** The last action performed in the plot pane can be disengaged and a removed trace can be brought back on the waveform viewer by using an F9 key.)

Remember: For drawing attention to a specific trace, keep it by double-clicking the wire, node, component, or pin (all other traces from the waveform viewer get deleted). Clicking the same voltage or current two times erases all other traces and the double-clicked trace remains plotted.

(B) Plotting by Manipulating the Waveform Viewer Window

i. **Using the View -> Visible Traces Options from the schematic menu bar**

After running a simulation and having a blank waveform viewer on the schematic, use the **View** -> **Visible Traces** schematic menu options. It allows the user to select an initial trace name for plotting from the options under the **Select Visible Waveforms** dialog window. It also gives random access to a complete list of data whose traces can be plotted.

2.4.2 Notes and Annotating Options for the Plot Pane

After running a simulation, either left-click somewhere inside the waveform viewer window (on a graph window or plot pane) or maximize a plot pane or else click on the waveform viewer file icon next to the schematic file icon in order to make the waveform viewer menu bar active. This action brings the **Plot Settings** waveform menu on the desktop menu bar while all the schematic tools become inactive. The user can click on the **Plot Settings** trace/waveform menu Plot Settings to review its sub-menu options (Place Text, Draw Arrow, Draw line, etc.) for editing a traced plot. Here, the drafted schematic is saved as diodedcsweep.

For example, under the **Plot Settings** plot menu, move a cursor on the **Notes & Annotations** option Notes & Annotations to choose the Draw Line sub-option for drawing a line (dotted by default) on the plot. Afterward, the user can right-click on the line drawn to edit the line annotation and change its line style from dotted to solid for having a more informative line on the trace.

Also, we can right-tick the Grid check box in the Plot Settings trace menu so that grid points appear on the plot pane. A text can also be easily placed on the plot.

To conclude, annotations for a plot along with the move and drag can be found by going to the Plot Settings -> Notes & Annotations options. It is also possible to save the annotations via the Save Plot Settings (available under the Plot Settings plot menu) sub-menu option.

Remember: When the waveform viewer menu bar (after clicking the waveform viewer file icon diodedcsweep) is active, all the schematic tools become inactive.

To make the schematic tools active, click on the schematic file icon or button diodedcsweep.

2.4.3 Adding Traces after Running a Simulation

- Using the **Add traces to Plot** Dialog Window:

Left-clicking the **Add trace** option Add trace Ctrl+A under the **Plot Settings** waveform viewer menu opens up the dialog window called **Add Traces to Plot.** The dialog window further allows adding a trace by clicking on the desired name of the trace. This action does not replace the already plotted current trace, instead, it keeps on adding the newly selected traces to the older trace.

The **Add Traces to Plot** dialog window includes all the available data sources for a simulation and the text box to enter an expression to be traced. Therefore, the user can easily add an **expression**(s) on any trace for plotting by clicking on the names of the traces instead of typing out their names in an allowed way. Also, two identical (one will be modified later) plots of a single data source can be entered by either typing in the expression or by clicking on the data source in the window twice.

Alternatively, after running a simulation, a trace or traces can be added manually by right-clicking somewhere inside the plot pane and selecting the **Add Traces** (see Fig. 2.5) option from the context menu.

2.4.4 Adding Supplementary Plot Panes (To Display Multiple Traces Separately)

- Using the **Context-based Plot** Menu:

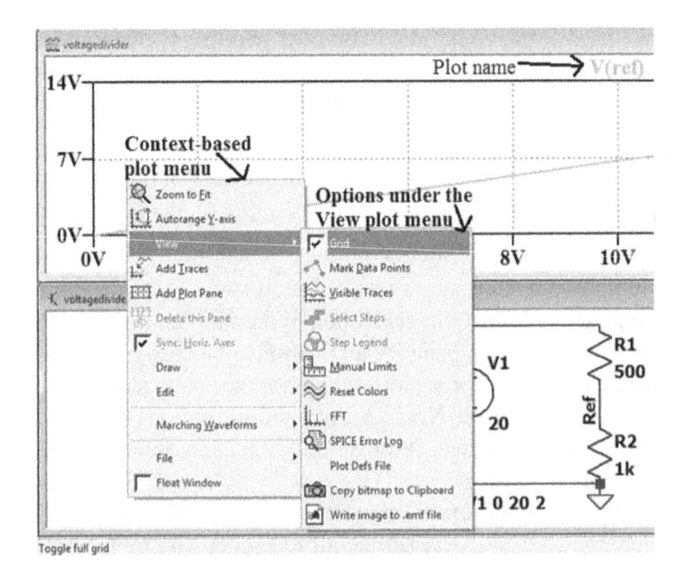

FIGURE 2.5 Adding Grid to the Plot Pane.

Inserting another plot pane to an existing one implies that another plot should have a separate vertical axis and optionally separate horizontal axis. This action assists the user in organizing information in a much sensible way by avoiding multiple scales on the same pane. This becomes important when plots look messy or we need to plot graphs that are at severely different scales from each other. To add a supplementary pane, right-click somewhere inside an existing plot pane and select the **Add Plot Pane** option Add Plot Pane from the context-based plot menu (see Fig. 2.5).

(**Note:** The traced graphs can be dragged from one plot pane to another when desired. To achieve this, left-click the trace name/label and hold a mouse button to drag and drop.)

2.4.5 Display Grid Points to the Waveform Viewer

In the waveform viewer window, right-click somewhere inside a plot pane and select the View -> Grid options. Right-tick the Grid check box so that when the schematic circuit is probed, the result is plotted with grid points on the plot pane.

Also, the user can select any additional feature such as the Zoom to Fit option (Fig. 2.5), Add Plot Pane, Draw, Edit, etc. A mathematical expression can also be entered using the Visible Traces and Add Traces options and bringing the respective dialog window on the screen.

The user can also assign a different color to a trace or completely delete a trace from the plot pane by hovering over the plot name assigned automatically by the simulator and right-clicking on it.

2.4.6 Automatic Display of Abscissa and Ordinate Values

It is to be noted that there exists a **mouse** cursor **readout** display or status bar at the bottom left corner of the schematic window, known as the Information Bar. When the user moves a mouse over the waveform window on any point, the mouse position can be read on the schematic Information Bar (or status bar). Also, by dragging a mouse, the size of a box can be displayed on the Information Bar to measure differences.

Therefore, placing a plus cursor on a data point on the trace which is to be measured, the x and y measurement values (where a mouse cursor points to) get displayed in the Information Bar as shown below in Fig. 2.6.

2.4.7 Doing Graphical Measurements and Mathematical Operations

a. **Integrating the trace for an average and RMS voltage/current:** The LTspice XVII waveform viewer can integrate a trace to produce an average or RMS value over the displayed region or for the selected abscissa range. A trace in the waveform viewer can be integrated by zooming in to a region of interest, and then while holding down a Ctrl key, left-click the trace label to be integrated. This action brings a new window on the screen displaying average and RMS

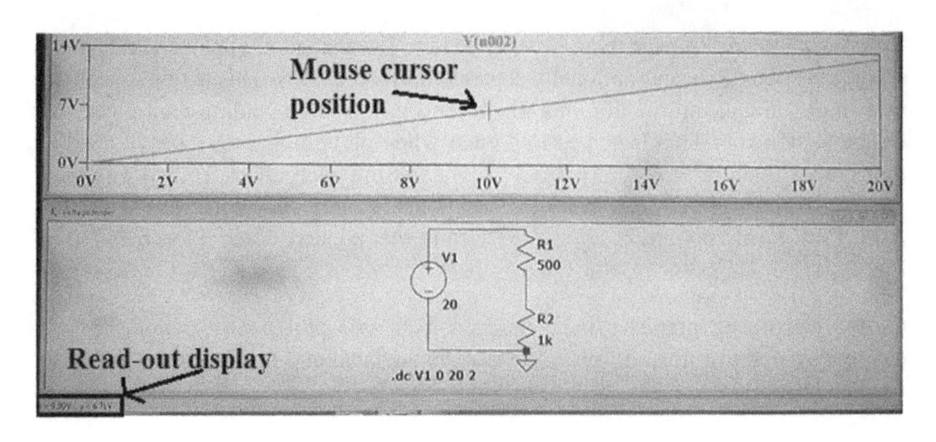

FIGURE 2.6 Reading x- and y-Axis Values from the Information Bar.

values by calculating a definite integral of the trace with a lower limit of integration to be zero and an upper limit being the recent time in the simulation.

b. **Instantaneous power:** To measure an instantaneous power dissipated or supplied by the component, hold an Alt key and left-click over the component symbol (a pointer changes into a thermometer, and the result is displayed in the default unit watts) whose instantaneous power is needed to be displayed.
c. **Average power:** While holding down a Ctrl key, left-click on the trace label/name of the power dissipation waveform.

Using the Expression Editor Window for Editing an Existing Trace:

a. **To enter an algebraic expression for editing an existing trace:** To plot an arithmetic function of node voltages and/or branch currents, move a cursor on the already plotted trace name (assigned automatically to all plots in the waveform viewer window) and right-click when a pointing finger cursor appears. The default name generally specifies the parameter type plotted along with the name of the nodes at/across or components through which measurements are taken) and is present at the top of the plot windowpane. Right-clicking on the trace name brings the **Expression Editor** window that allows the traced expression to be manually modified, the default color to be changed, the numbered cursor or cursors to be attached or the existing trace to be deleted completely from the plot pane. Thus, the user can enter a mathematical expression (containing valid function names and operators to be applied to the traced parameter and plot the results.

   Alternatively, a mathematical expression employing any one or all of the signals present in the circuit can be entered by first clicking on the (.raw) and making the waveform viewer window the active window. Afterward, using the Visible Traces and Add trace sub-menu options under the Plot Settings waveform menu, the desired expression can be added in the Expression Editor box. Alternatively, both the options Visible Traces (in the

View context-based menu) and Add Traces can also be accessed by right-clicking somewhere inside a plot pane (see Fig. 2.5).

b. **Applying the built-in functions on a waveform data:** The waveform viewer window allows several built-in mathematical functions for plotting so that the user can manually enter/edit functions. Table 2.3 provides a list of the available functions for real data.

The following functions shown in Table 2.4 are available for complex data:

The following operations shown in Table 2.5 are grouped in **reverse** order of **precedence** of evaluation:

The global variables and constants that are internally defined as **keywords** are shown in Table 2.6 below.

The keyword time is implicit when plotting a transient analysis waveform data, whereas the freq is understood when simulating an AC analysis. Also, the default units for the available data sources (case insensitive) include V, A, Ohm, W (watt), J (joule), s (second), Hz, and deg.

To conclude, mathematical operations can be performed on waveform data in three ways:

- Entering an expression of available simulation data traces to plot the expressions of traces,
- Determining an average or RMS of a plotted graph or trace, and
- Computing Fourier Transform of a data-trace.

2.4.8 User-Defined Functions (dot func command)

- **Using the Func Dot Command (.func directive):**

The users can define their own functions to be used with the traces plotted using the already defined functions, variables, and constants. Place the .func directive statement on the schematic page (or define a function by choosing the Plot Settings -> Edit Plot Defs File) with the following syntax:

```
.func function_name(arguments) {mathematical expressions}
```

For example, to define the function that converts radians to hertz, place the directive statement as shown below:

```
.func frequency(omega){omega/(2*pi)}
```

which takes a value of an angular frequency (omega) in rad/sec and obtains its corresponding value in hertz.

To use the function in a parametric expression, substitute the given variable, say omega with the required value, and then paste frequency(omega) into wherever it is needed, say the value/expression text field of the required global variable f as shown below (Fig. 2.7):

TABLE 2.3
The Built-in Function Names and their Description

Function Name	Description
abs(x)	An absolute value of the x
acos(x)	A real part of an arc cosine of the x, e.g., acos(−5) returns 3.14159, not 3.14159 + 2.29243i
arccos(x)	A synonym of the acos()
acosh(x)	A real part of the arc hyperbolic cosine of the x, e.g. acosh(.5) gives 0 and not 1.0472i
asin(x)	A real part of an arc sine of the x, e.g. asin(−5) outputs −1.57080 and not −1.57080 + 2.29243i
arcsin(x)	A synonym of the asin()
asinh(x)	An arc hyperbolic sine
atan(x)	An arc tangent of the x
arctan(x)	A synonym for the atan()
atan2(y,x)	A four quadrant arc tangent of the y/x
atanh(x)	An arc hyperbolic tangent
buf(x)	Outputs 1 if the x > .5, else returns 0
cbrt(x)	Cube root of the x
ceil(x)	An integer equal to or greater than the x
cos(x)	A cosine of the x in radians
cosh(x)	A hyperbolic cosine of the x
ddt(x)	Returns a time derivative of the x
exp(x)	e to a power of the x
fabs(x)	Same as the abs(x)
flat(x)	Generates random numbers between the variables -x and x with a uniform distribution
floor(x)	An integer equal to or less than the x
gauss(x)	Random numbers from a Gaussian distribution with a sigma of the x.
hypot(y,x)	Hypotenuse i.e. sqrt(x*x+y*y) or sqrt(x**2+y**2)
idt(x[,ic[,a]])	Integrates the x, the optional initial condition ic resets if the a is true
if(x,y,z)	If the x > .5, then outputs y else z
int(x)	Converts the x into an integer
inv(x)	Returns 0 if x > .5, else 1
limit(x,y,z)	An intermediate value of the x, y and z i.e. equivalent to min(max(x,y),z)
ln(x)	Natural logarithm of the x
log(x)	An alternate syntax for the ln()
log10(x)	Base 10 logarithm
max(x,y)	Greater of the x or y
mc(x,y)	A random number between x*(1+y) and x*(1-y) with a uniform distribution.
min(x,y)	Smaller of the x or y
pow(x,y)	x**y (A real part of the x raised to the power y, zero for the negative x and fractional y, e.g. pow(−.5,1.5) returns 0 and not 0.353553i)
pwr(x,y)	abs(x)**y (An absolute value of the x to the power y)

(Continued)

TABLE 2.3 (Continued)

Function Name	Description
pwrs(x,y)	sgn(x)*abs(x)**y
rand(x)	A random number between 0 and 1 depending on the integer value of x
random(x)	Similar to the rand(), but smoothly transitions between values
round(x)	Nearest integer to the x
sgn(x)	Sign of the x i.e. it returns −1 for x < 0, 0 for x == 0 (the operator == implies exactly equal to and 1 for x > 0
sin(x)	Sine of the x
sinh(x)	A hyperbolic sine of the x
sqrt(x)	A real part of a square root of the x and zero for the negative x, e.g. sqrt(−1) returns 0 and not i
table(x,x1,y1...)	Interpolate the y(x) based on a lookup table given as the x-ordered set of pairs of points
tan(x)	Tangent of the x
tanh(x)	A hyperbolic tangent of the x
u(x)	Unit Step (outputs 1 if x > 0, else returns 0)
uramp(x)	Unit Ramp (outputs x if x > 0, else returns 0)

TABLE 2.4
Functions for Complex Data

re(x)	Returns a real part of a complex number (argument) x
im(x)	Returns an imaginary real part of a complex number x
ph(x)	Returns a complex number equal to the phase angle of an argument x
mag(x)	Returns a complex number equal to the magnitude of an argument x
db(x)	The complex number x magnitude in dB
conj(x)	Returns a complex conjugate of the argument x

```
.param omega = 2*pi   f= frequency(omega)
```

A function may be required in several places in a schematic or it is useful in several different schematics to avoid copying and pasting a complicated expression as a block of text each time it is needed.

Another example of using the .func directive is shown below (Fig. 2.8):

TABLE 2.5
The Valid Operators and their Description

<table>
<tr><th>Operator</th><th>Description</th></tr>
<tr><td>&</td><td>Converts expressions (containing real data only) on either side of the operator to Boolean, and then AND</td></tr>
<tr><td>|</td><td>Converts adjacent expressions (containing real data only) on either side of the operator to Boolean, and then OR</td></tr>
<tr><td>^(caret character)</td><td>Converts expressions (containing real data only) on both sides of the operator to Boolean, and then XOR. Also, implies exponentiation in the case of a complex data</td></tr>
<tr><td></td><td>For Boolean operations, true is 1 and false is 0. Boolean conversions return true if evaluates to greater than .5, else false</td></tr>
<tr><td></td><td>The Boolean XOR operator ^ represents exponentiation when used in a Laplace expression</td></tr>
<tr><td><</td><td>True if an expression on the left is less than an expression on the right, otherwise false</td></tr>
<tr><td>></td><td>Returns true (logical 1) if an expression on the left is greater than an expression on the right, otherwise false (logical 0)</td></tr>
<tr><td><=</td><td>True if an expression on the left is less than or equal to an expression on the right, otherwise false</td></tr>
<tr><td>>=</td><td>True if an expression on the left is greater than or equal to an expression on the right, otherwise false</td></tr>
<tr><td>==(double equal sign)</td><td>True if preceding and succeeding expressions are equal, otherwise false</td></tr>
<tr><td>+</td><td>Floating-point addition</td></tr>
<tr><td>-</td><td>Floating-point subtraction</td></tr>
<tr><td>*</td><td>Floating-point multiplication</td></tr>
<tr><td>/</td><td>Floating-point division</td></tr>
<tr><td>^</td><td>Exponentiation of complex data only</td></tr>
<tr><td>**</td><td>Raises a data on the left-hand side to the power of a data on the right-hand side, and returns an answer without displaying a complex notation i, e.g., 2**(1+i) returns 2 and not 2i</td></tr>
<tr><td>! or ~</td><td>Converts a succeeding expression to Boolean, and then inverts</td></tr>
</table>

2.4.9 Attaching the Numbered Cursors (To Read Data Values)

LTspice XVII waveform plotting routine includes two numbered cursors. A single cursor can be activated simply by left-clicking on a label of the traced plot so that a cursor box (the Cursor readout display window) appears displaying the horizontal and vertical coordinates of the waveform point that initially exists under the cursor. Alternatively, the user can right-click on the desired plot label and select the cursors from the Attached Cursor drop-down menu (the options are none, 1st, 2nd, or 1st & 2nd) at the top of the Expression Editor window.

Thus, the user can attach the LTspice XVII numbered cursor or cursors for finding the exact x and y values of a specified data point in the floating cursor

TABLE 2.6
The Keywords and their Pre-defined Values

Name	Value
e	2.7182818284590452354
pi	3.14159265358979323846
k (Boltzmann constant)	1.3806503e-23
q (charge of an electron	1.602176462e-19
i (imaginary unit) (complex data only)	sqrt(−1)
c (speed of light in meters/sec)	2.99792 e+08
freq (frequency in Hertz)	variable
omega (in rad/sec) (complex data only)	variable
time (in seconds)(real data only)	variable
kelvin (absolute zero in degrees)	−2.73150 e+02

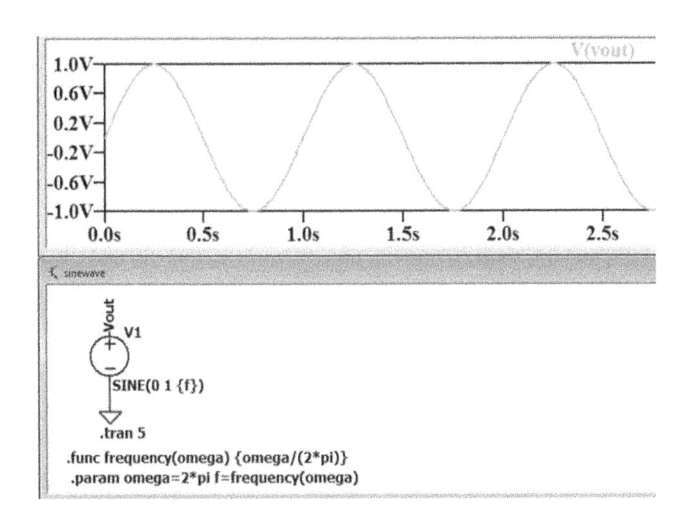

FIGURE 2.7 Converting Radians into Hertz with a User-Defined Function.

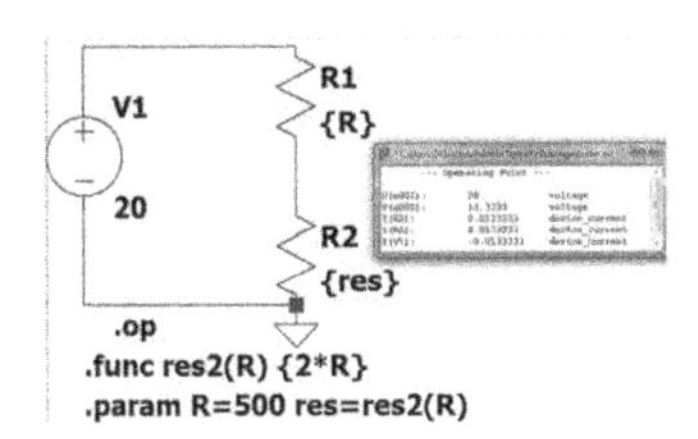

FIGURE 2.8 Using the .func Directive Statement for Relating the Two Resistances.

information window (or a readout display window). The cursor information window can be moved to an appropriate location by left-clicking on it and dragging a mouse. Click on the name of the plot or trace to pop-up the **Expression Editor** dialog window having the **Attached Cursor** drop-down menu.

The drop-down menu enables the user to select between Cursor 1 or Cursor 2 or both. The numbered Cursor 1 (or the Cursor 2) when attached sets a cursor cross-bar to display horizontal and vertical axis information about the point (at which the cursor is placed in the beginning) in a separate floating window.

The numbered cursor (the Cursor 1 or the Cursor 2) can be easily moved back and forth (by single-clicking and dragging) through the data points by dragging with a mouse or using the cursor keys. First, move a mouse pointer over the cross-bar so that a number 1 (in yellow color) appears, and then drag a mouse while left-clicking on the number to move the cursor along with the graph and position it to display the desired data point horizontal and vertical values.

If the user selects to attach two cursors by choosing the 1st & 2nd option, it then allows automatic math to be done, for example, gives a difference between them in both the dimensions.

(**Note:** To remove the cursor(s), simply close the floating information cursor window.)

3 Control Panel Settings

3.1 SCHEMATIC GRAPHICAL USER INTERFACE SETTINGS WITH CONTROL PANEL

The user can make various useful settings or adjustments to change the appearance of the schematic window using the software's Control Panel dialog window.

3.1.1 Preliminary Schematic Diagram Settings

The schematic diagrams are limited to one page and have specific preliminary settings. By default, an LTspice XVII circuit diagram constructed on the new schematic is color-coded so that component parts, component text, comment text, and wires can be easily distinguished and understood as shown in Fig. 3.1.

To change the default schematic graphical user interface settings, open the Control Panel window either by clicking on the Control Panel (hammer icon) button on the schematic toolbar or selecting the Tools -> Control Panel options from the pull-down schematic menu bar. By clicking on the **Drafting Options** tab, the schematic initial settings can be changed. To change the default color scheme, select the **Color Scheme** button so that the **Color Palette Editor** can be accessed.

In the **Color Palette Editor** window, ensure that the **Schematic** tab (see Fig. 3.2) is active. Here, the color of the items available in the pull-down menu next to the **Selected Item** field such as component body, wire, graphic flag, component text, etc. can be changed by changing the RGB values of the selected item. Also, moving a cursor on the item and left-clicking automatically displays the item name in the **Selected Item** field.

- **Changing the Schematic Background Color to White**

In the **Color Palette Editor**, click on a downward pointing black triangle (drop-down arrow) next to the Selected Item field to have the pull-down menu and choose the option **Background** (see Fig. 3.2) from the drop-down menu. Specific color for all the available shown items can be picked by moving the Red, Green, and Blue sliders, and then clicking the **Apply** and **OK**. It is always preferable to set the schematic background color to white (RGB values are R = 255, G = 255 and B = 255) for more clarity.

To change the schematic background color to white, move the three Red, Green, and Blue (RGB) sliders to the rightmost corner so that they are at 255. This is desired to make things look more comprehensible and apparent. Alternatively, the

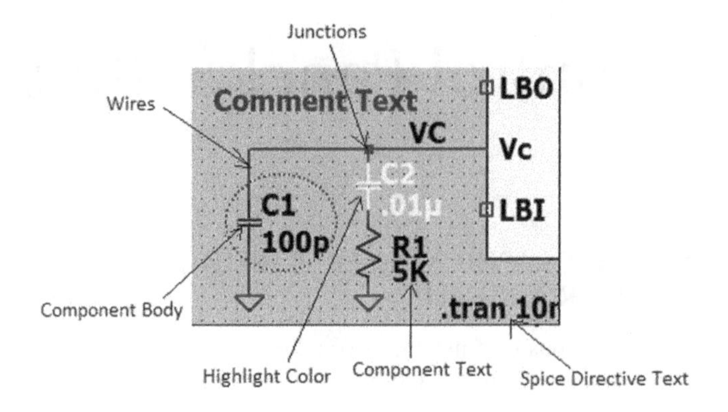

FIGURE 3.1 The Default Circuit Diagram Settings.

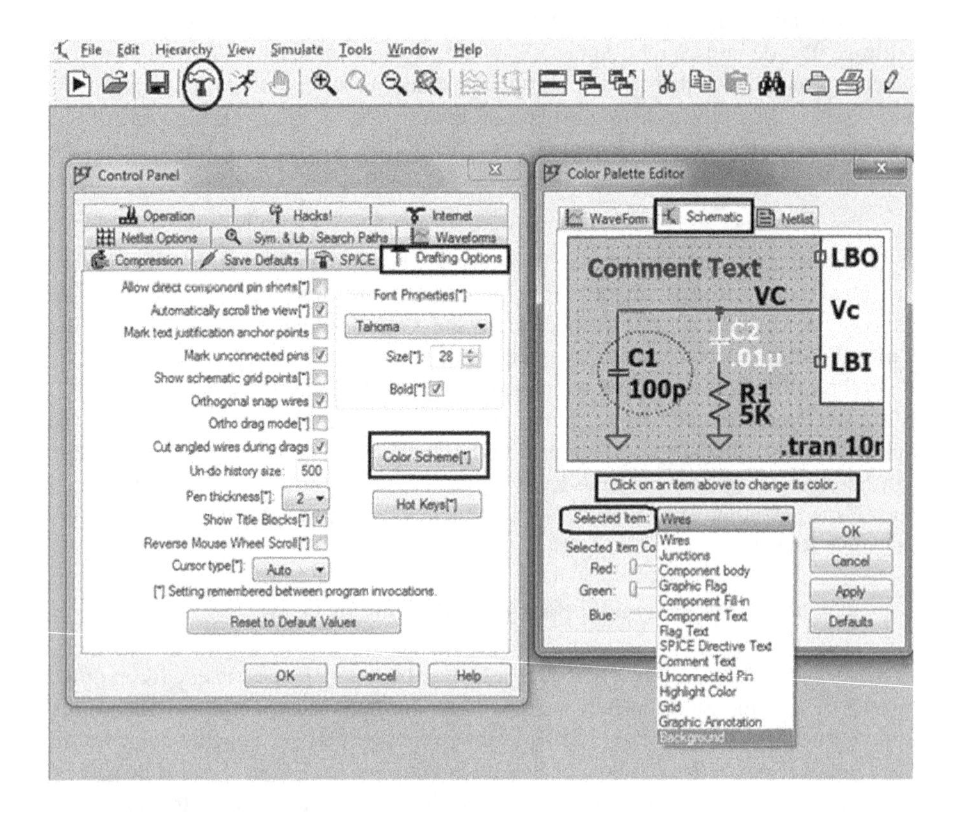

FIGURE 3.2 The Schematic Tab under the Color Palette Editor.

RGB values can be modified with the numbers (typing R, G, and B values in the respective boxes instead of moving the slider bars. The RGB values of frequently liked colors are shown in Table 3.1.

TABLE 3.1
Different Colors and their RGB Values

Color	Red (R)	Green (G)	Blue (B)
White	255	255	255
Yellow	255	255	0
Orange	255	165	0
Gray	128	128	128
Light Gray	211	211	211
Light Blue	173	216	230
Green Yellow	173	255	47
Ocher	195	147	67
Pink	255	192	203
Deep Pink	255	20	147
Blue Green	0	164	151
Red	255	0	0
Dark Green	0	128	0
Purple	128	0	128
Brown	165	42	42
Blue	0	0	255
Navy	0	0	128
Black	0	0	0

After moving the RGB sliders to the rightmost corner (i.e., at 255 RGB value), clicking the **Apply** and **OK**. The following window appears (Fig. 3.3):

Click the OK for inserting the changed setting.

Now, click on the New Schematic icon to open a schematic window with a white background as shown below (Fig. 3.4).

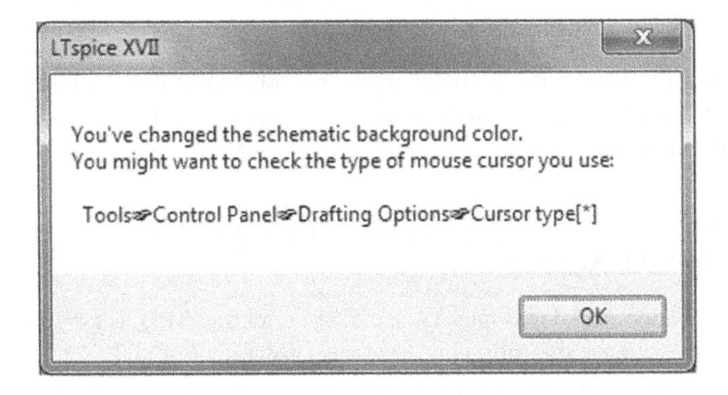

FIGURE 3.3 The Schematic Background Color Is Turned To White.

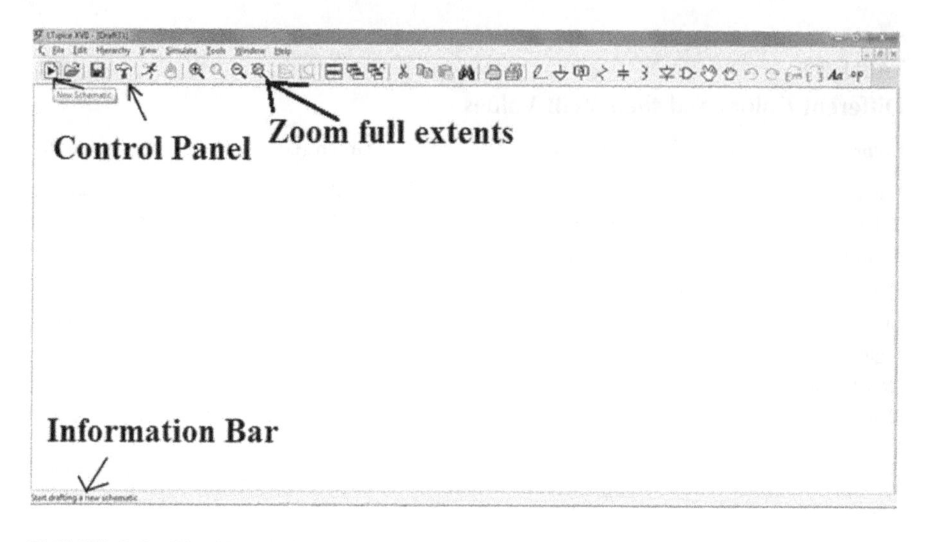

FIGURE 3.4 The New Schematic with a White Background.

3.2 GRID DISPLAY AND FONT PROPERTIES ON NEW SCHEMATIC

To display grid points on the empty new schematic, click the **Drafting Options** tab (Fig. 3.5) and right-tick the **Show schematic grid points** check box. Also, to return to the default settings (in case of any mistake done), left-click on the **Reset to Default Values** button. Here, the initial font properties of the text on the schematic can also be changed.

3.3 CHANGING SCHEMATIC BACKGROUND COLOR TO WHITE

The default color of the empty schematic is grey which can be changed easily for a better look. To modify the background color, go to the Tools -> Color Preferences desktop/schematic menu bar as shown in Fig. 3.6.

This also opens up the **Color Palette Editor** window to change the initial schematic diagram color settings.

(**Note:** When the user drifts a cursor over any schematic tool button, a brief description of the icon appears such as moving a cursor on the schematic icon ▶ displays that it is meant for creating the new schematic.)

3.4 MODIFICATION OF SIMULATION SPEED AND INTERNAL ACCURACY

To set a simulation speed and internal accuracy, click the **SPICE** tab in the Control Panel window to open the following fields as shown in Fig. 3.7.

The user may choose the **Alternate** option in the Engine -> Solver sub-field for high internal accuracy, provided there is enough processing power of a PC. The default Normal Solver option provides simulation speed equal to 2 times and

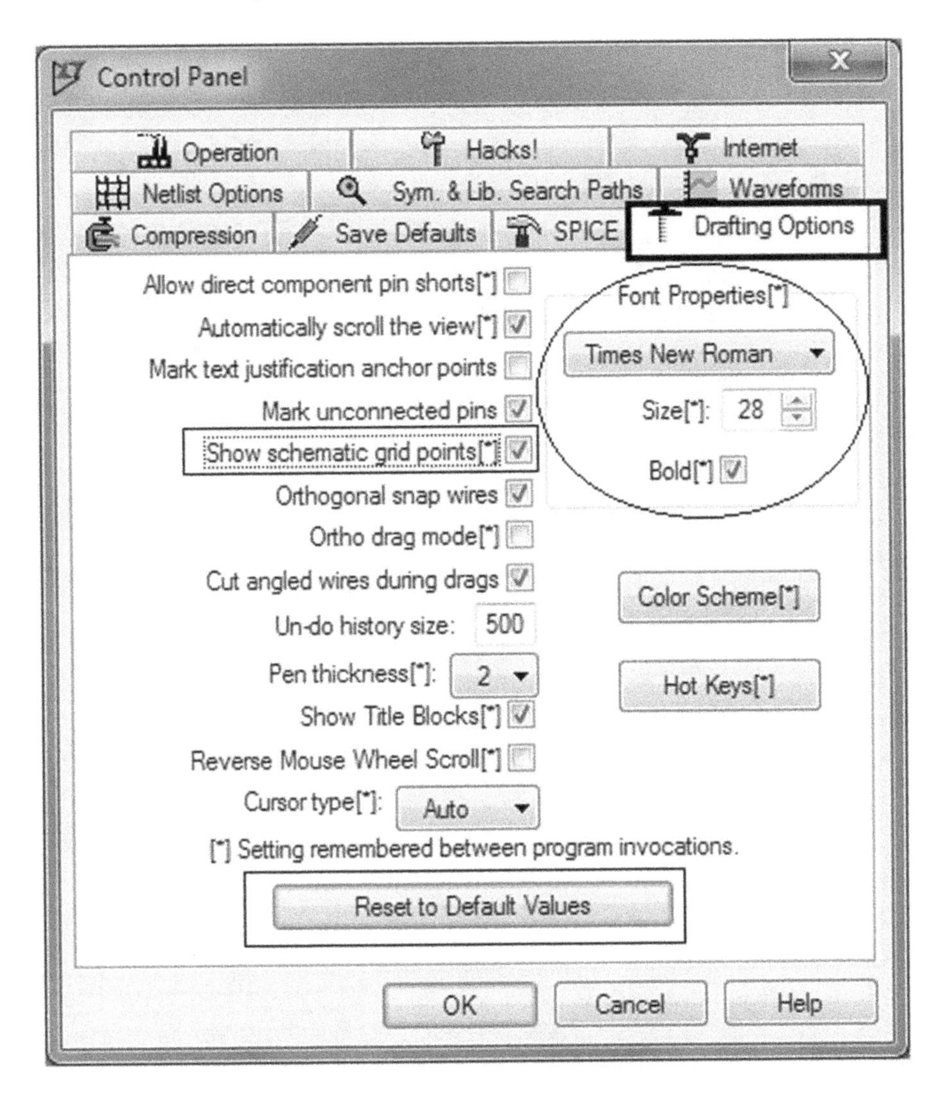

FIGURE 3.5 The Screen Displaying the Fields of the Drafting Options Tab.

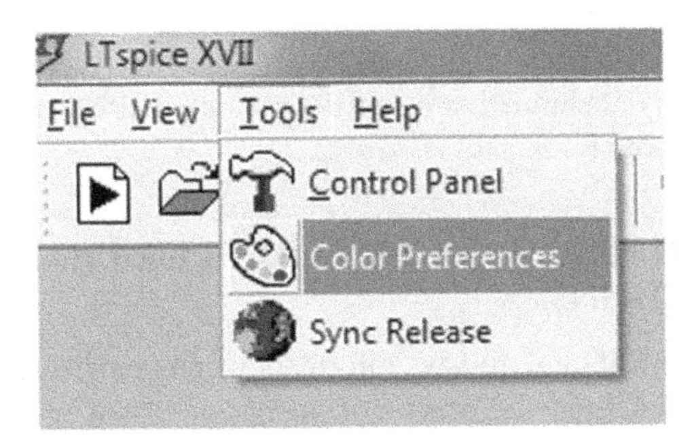

FIGURE 3.6 The Start-Up Window (the Tools Schematic Menu).

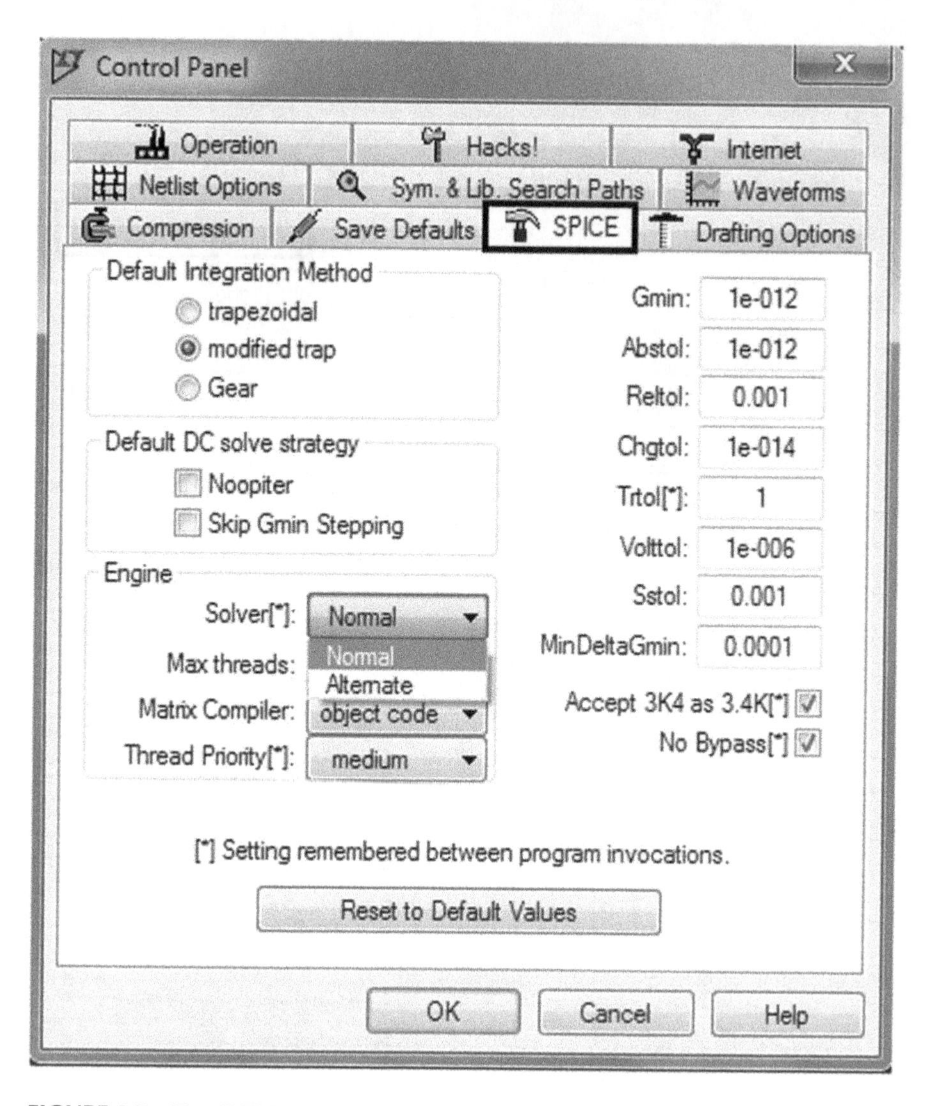

FIGURE 3.7 The SPICE Tab under the Control Panel Screen.

internal accuracy 1/1000 as compared to the option Alternate. Whereas, the **Alternate** option gives simulation speed ½ and internal accuracy 1000 times as compared to the Normal.

3.5 SELECTING NEW SHORTCUTS (KEYS OR CHARACTER COMBINATIONS)

The default shortcut F (function) keys information can be obtained by choosing the Help -> Help Topics from the schematic menu, and then selecting (by double-clicking) the Schematic Capture option by double-clicking the LTspice XVII icon under the Contents tab menu as shown in Fig. 3.8.

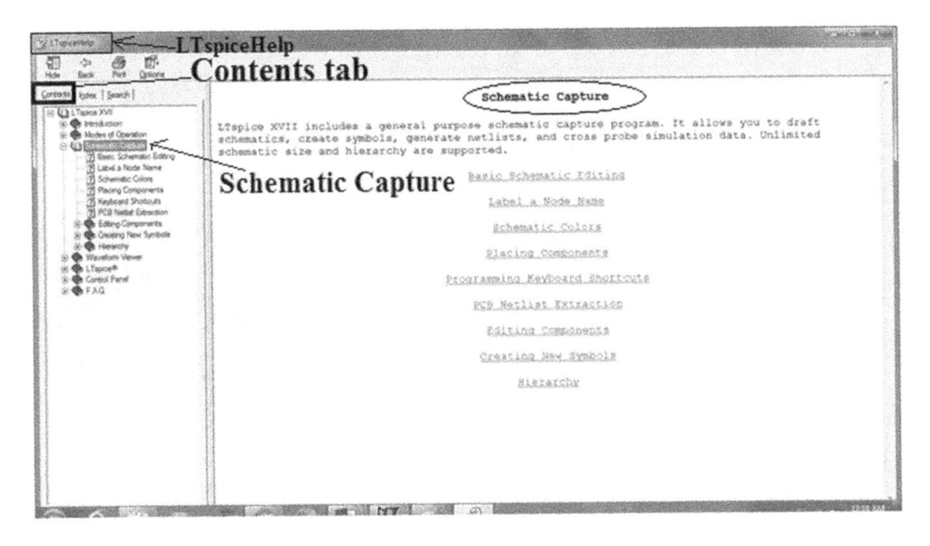

FIGURE 3.8 The Available Options under the Contents Tab of the Help Section.

Keyboard Shortcut Map

Click an item below and then press the new keyboard combination for the new short cut.

Schematic | Symbol | WaveForm | Netlist

Place Component:	F2	Place Diode:	D
Draw Wire Mode:	F3	Draw Lines:	(none)
Place Netname:	F4	Draw Rectangles:	(none)
Delete Mode:	F5	Draw Circles:	(none)
Duplicate Mode:	F6	Draw Arcs:	(none)
Move Mode:	F7	Reset T=0:	0
Drag Mode:	F8	Schematic Grid:	Ctrl+G
Undo:	F9	Rotate:	Ctrl+R
Redo:	Shift+F9	Mirror:	Ctrl+E
Place Comment Text:	T	Zoom Area:	Ctrl+Z
Place SPICE Directive:	S	Zoom Back:	Ctrl+B
Unconn. Pin Marks:	U	Halt Simulation:	Ctrl+H
Text Anchor Marks:	A	Run Simulation:	(none)
Place Ground:	G	Place BUS Tap:	(none)
Place Resistor:	R	Zoom to Fit:	Space
Place Capacitor:	C	View SPICE Error Log:	Ctrl+L
Place Inductor:	L		

Cancel | Reset to Default Values | OK

FIGURE 3.9 The Keyboard Shortcut Map Display Window.

The default keys can also be modified to characters as follows:

Click the **Drafting Options** tab (see Fig. 3.5) to open all its fields and buttons. Click the button **Hot Keys** [Hot Keys(*)] to pop up the following **Keyboard Shortcut Map** display window (Fig. 3.9).

The user can change keyboard shortcuts by clicking on the function to be performed, and then pressing a new keyboard character combination so that they can be easily remembered.

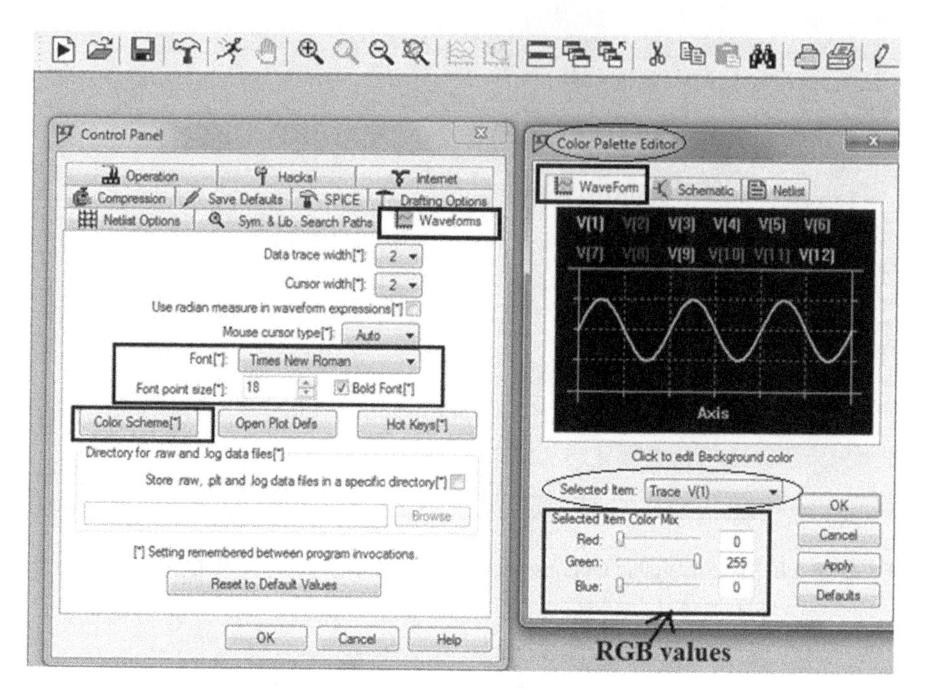

FIGURE 3.10 Editing the Traced Plot Settings.

3.6 EDITING INITIAL PLOT PANE SETTINGS

By clicking on the **Waveforms** tab in the **Control Panel** dialog window, the initial font properties of the text that appears on the trace in the waveform viewer window can be changed easily (Fig. 3.10). Here, the waveform viewer window color settings can also be changed selecting the **Color Scheme** button (Fig. 3.10). The action further opens the Color Palette Editor window where the desired color can be selected for the specified trace.

The user can change the color of the waveform plotted/traced (or any other item such as axis color, trace background color, and grid color) by first selecting it from the pull-down menu next to the Selected Item field in the Color Palette Editor to make traces more readable and acceptable. Now, set a recommended color for the desired trace (or any other selected item) by changing the **RGB** values (moving the Red, Green, and Blue sliders or typing their values in the text boxes next to them).

Always click the **OK** to insert the changes made.

Alternatively, click the **Tools -> Color Preferences** in the schematic menu bar to bring the Color Palette Editor.

To Display a Waveform Trace alongside the Schematic Diagram, click the **Operation** tab in the Control Panel window and choose the **Vert** from the

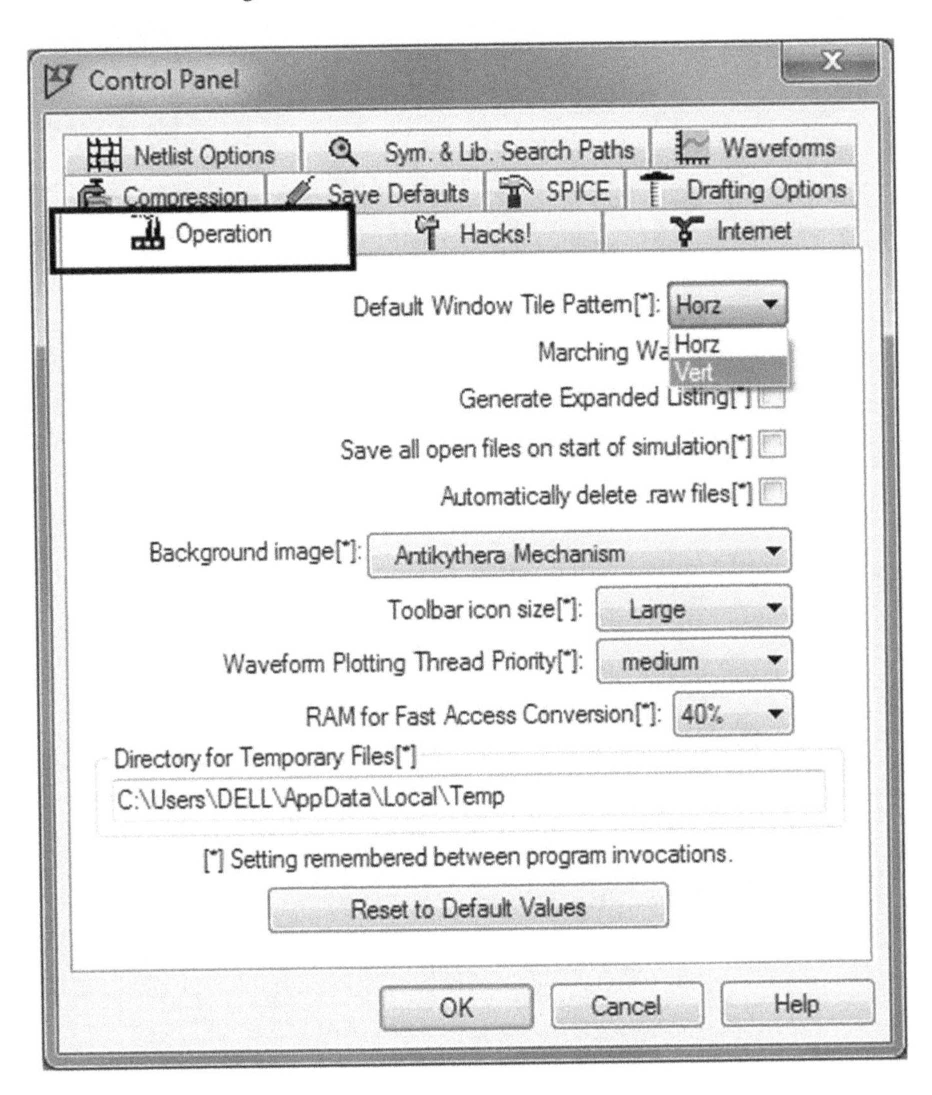

FIGURE 3.11 Displaying a Trace alongside the Schematic Diagram.

drop-down menu next to the Default Window Tile Pattern field as shown in Fig. 3.11.

Click the OK. Thus, the user can make several modifications to the default schematic graphical user interface settings according to preferences.

4 DC Bias and DC Sweep Simulations

4.1 INTRODUCTION

- **DC Operating Point Analysis**

It calculates a static behavior of the drafted circuit when a DC voltage or current is applied to it. The results of the analysis include bias point or quiescent point measurements at the single-source value. In most cases, the results of the analysis are intermediate values for further analysis.

- **DC Sweep Analysis**

Whereas, a DC sweep analysis is used to calculate the circuit's bias point over a range of voltage/current values. In the analysis, the simulator steps a parameter value of the specified independent voltage or current source over a user-specified range and performs an operating point analysis at each value. The analysis is useful for examining I-V (current-voltage) curves of the component models for verification and computing a DC transfer function of an amplifier.

4.2 DC VOLTAGE SOURCE WITH RESISTOR (AN OPERATING POINT ANALYSIS)

To study the behavior of a resistor when a DC voltage or current is applied to it, perform a DC operating point analysis (also referred to as a bias point or quiescent point or simply Q-point analysis). Start a simulation by first drafting the required circuit.

4.2.1 Simulating A Simple Circuit (DC Source Connected across a Resistor)

Steps Involved:

i. Construct the Circuit Schematic: Start LTspice XVII and open the new schematic window. Access the standard models provided within the simulator program to select appropriate circuit elements and place them on the schematic. Connect the circuit elements together through wiring using the Wire schematic tool in order to draft a virtual breadboard of the circuit as shown in Fig. 4.1.
   - To paste the given circuit in a document, go to the **Tools -> Copy bitmap to Clipboard** schematic menu options so that the given circuit appears as shown in Fig. 4.2.

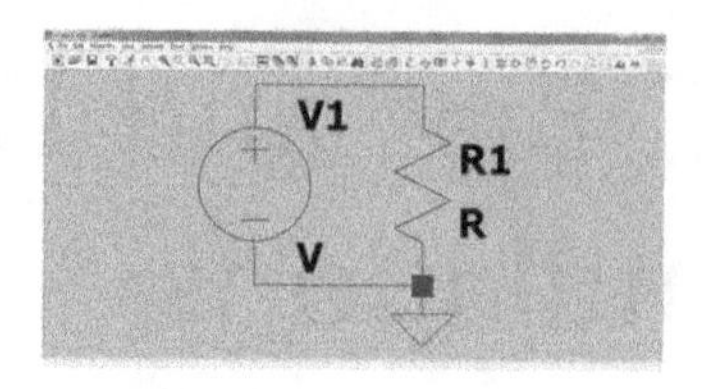

FIGURE 4.1 The Circuit Schematic.

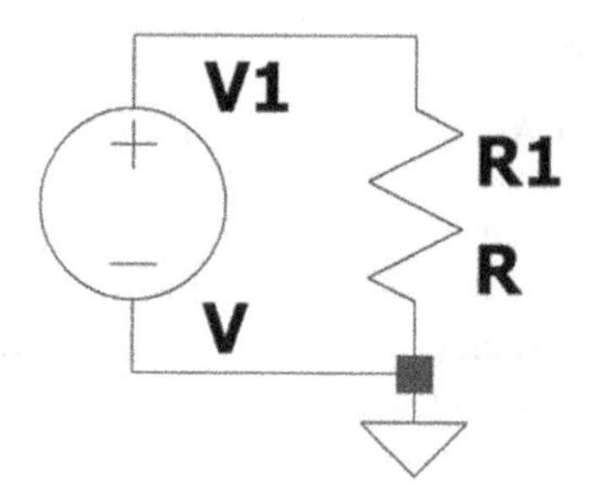

FIGURE 4.2 The Circuit Schematic Inserted in MS-Word.

- Changing the **Schematic Background** Color to **White:** Here, a white color is chosen for the background of the schematic diagrams instead of grey (the default). This is achieved as follows:

i. First, select the **Control Panel** schematic tool, and then click on the **Waveforms** tab to open its available fields and menu buttons and select the **Color Scheme** button. This opens the **Color Palette Editor**. Otherwise, go to the schematic toolbar and click the **Tools** -> **Color Preferences** options to open the **Color Palette Editor** dialog window.
In the already opened **Color Palette Editor** dialog window, click on the **Schematic** tab, and then next to the **Select Items** field box choose the **Background** option from the drop-down menu. Left-click on each of the slider (placed originally at the default position) under the Selected Item Color Mix field option and drag (with a left mouse button pressed) the Red, Green and Blue sliders one by one to the right-most place to set the RGB values to 255 so that the simulator can pick a white color for the schematic background.

(**Note:** Click on the Component icon from the schematic toolbar to open the component window with a wide range of common components and folders containing additional types of components and scroll to the far right column to select the voltage source model. Clicking on the **Component** symbol that looks like an AND gate brings up the dialog box which allows the user to browse components and folders to preview the symbol database.)

Remember: Any circuit drafted in LTspice XVII must have a ground node (global common node 0) before it can be simulated. Select the Ground (or reference node) icon that looks like a triangle from the schematic toolbar and place it on the circuit at an appropriate spot (usually near the bottom) of the circuit schematic. Press a right-key or an Esc key to exit the ground node (GND) selection.

IMPORTANT: The simulator cannot run the given circuit schematic in the current form with the components having non-numeric values. The voltage source name is V1 and its value is currently V volts. The resistor label is R1 and its value is currently R ohms.

ii. Assign parametric values to the components so as to make the simulator run an analysis and display the results.

- Setting the Independent Voltage Source Parameter Value

The Schematic Capture provides three ways to edit parameter values of the components and assign new numeric values by replacing the default values in the component properties editor window.

- **The Assisted Mode:** In this mode, the user right-clicks on the body of the component to invoke its specific GUI that assists in editing the component value. It's useful when the user wants to pick a semiconductor from the database of modeled devices or is unsure of the SPICE syntax trying to accomplish, such as entering a piecewise linear data for the current source.

Right-click on the component body, that is, the voltage source V1 to open its attributes/properties editor window as shown in Fig. 4.3.

In the given component editor window, press the **Advanced** button to open up the component menu dialog window which allows choosing among a DC voltage source, AC voltage source, pulse, sine wave, exponential, single frequency FM, or piece-wise linear waveform.

- **The Expert Mode:** In this mode, the user simply points a cursor at the default **value** text on the component symbol (its value editor window appears with a different GUI), then right-click and type in the desired numeric value for the component parameter. When the user moves a cross-hair pointing cursor near to the value text, it turns into a text caret so that the value can be edited.

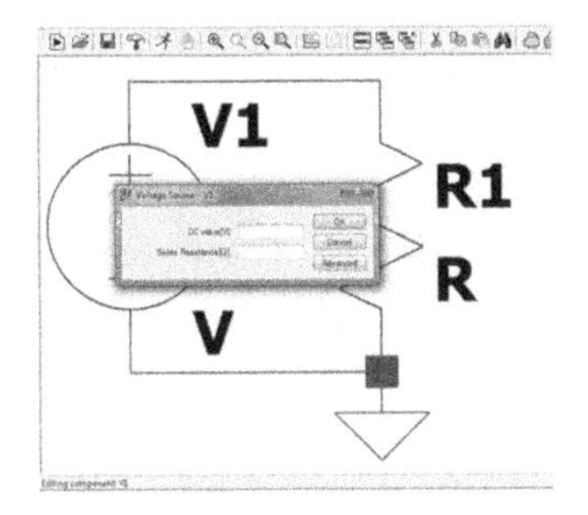

FIGURE 4.3 The Independent Voltage Source Properties Editor Window (using the Assisted Mode).

Thus, right-clicking on the value text V of the voltage source modelV1 brings up the component's value editor window having a different GUI from the window shown in Fig. 4.4.

- Setting the Resistor R1 Resistance Value
- **The Super Expert Mode:** In this mode, the user first points at the component symbol body, and then right-click while holding down a Ctrl key to open a more comprehensive editor window named the **Component Attribute Editor** window with a GUI as shown in Fig. 4.5 through which the user can control every available attribute of the component and edit the contents and visibility.

(**Note:** There is a check box next to each field which indicates that the entered field value should be visible on the schematic.)

- **The Assisted Mode:** Bring a cursor on the component body, that is, on the resistor model R1 symbol and right-click when a pointing finger appears as shown in Fig. 4.6.

The action opens its attributes editor window (the Resistor – R1) as shown in Fig. 4.7.

- **An expert mode:** Now, right-click on the value text R of the resistance model R1 (Fig. 4.8):

A value/property editor window (the **Enter new Value for R1**) appears (see Fig. 4.9) with a different GUI from the window with GUI shown in Fig. 4.7.

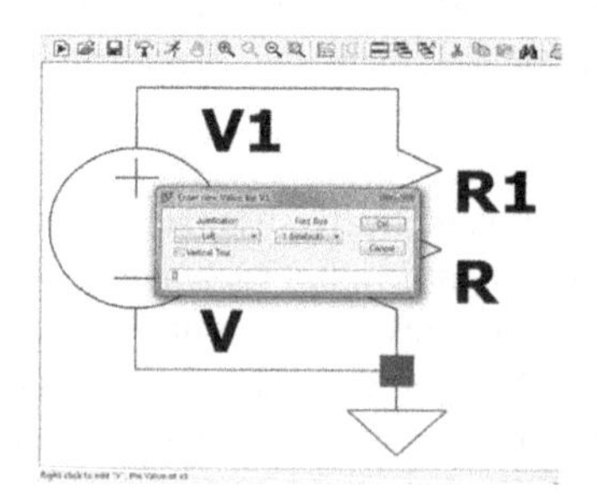

FIGURE 4.4 The Independent Voltage Source Properties Editor Window (from an Expert Mode).

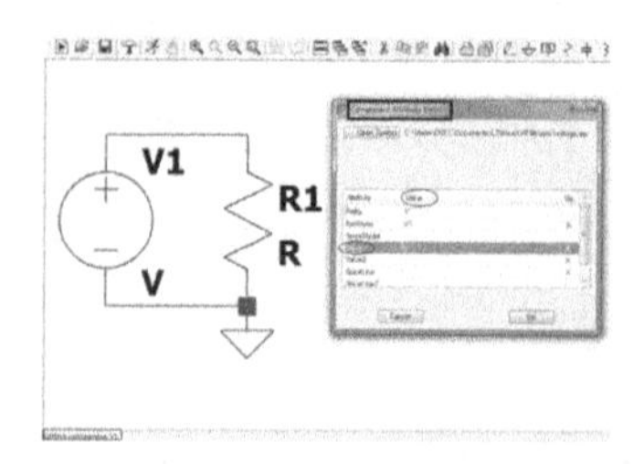

FIGURE 4.5 The Component Attribute Editor Window for the Resistor model R1.

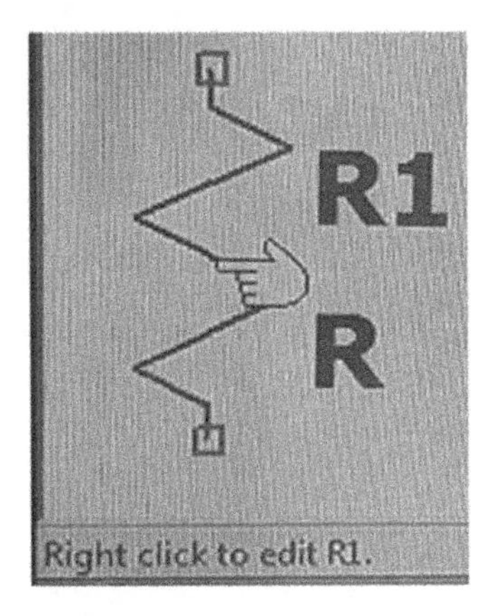

FIGURE 4.6 The Resistor Model R1 Symbol.

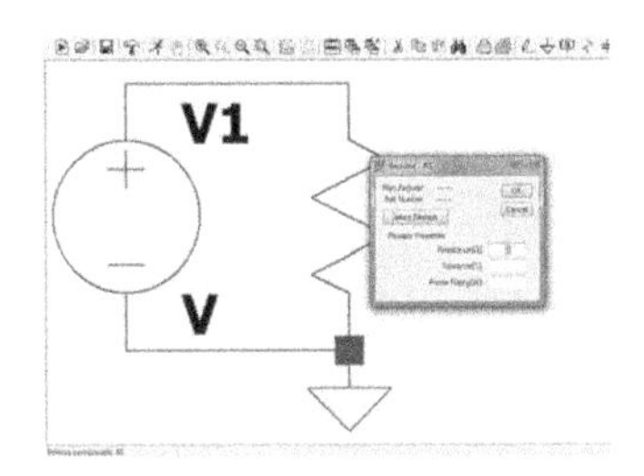

FIGURE 4.7 The Resistor Component Properties Editor Window (using the Assisted Mode).

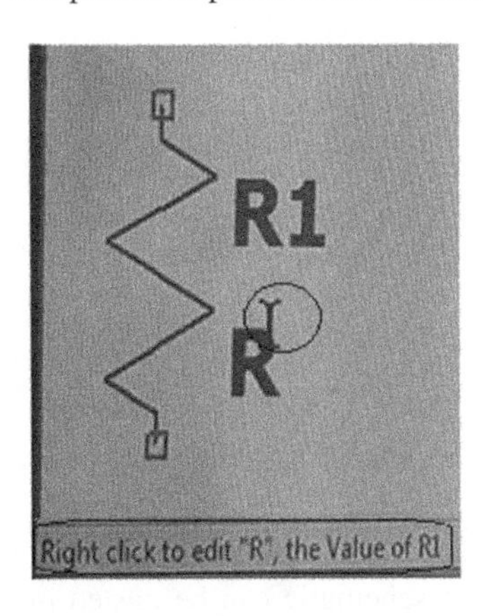

FIGURE 4.8 The Resistor Model Value Text R.

iii. Set a DC value of 10 V for the voltage source V1 and a resistance value of 0.9 kΩ (kiloohm) for the R1. We can enter either the numeric value as 0.9k or 0k9 for a resistance parameter.
iv. Save the circuit schematic with a representative name so that it can be accessed again. Go to the File -> Save As, and then type a desired name, say **simplecircuit** as shown in Fig. 4.10. Click the Save button to finish saving the schematic file with a .asc extension.

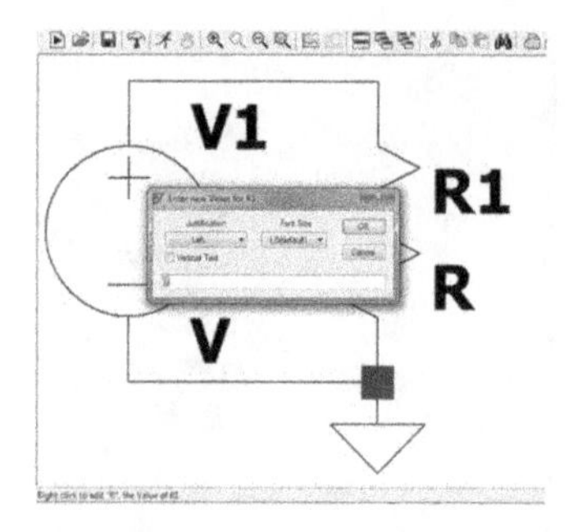

FIGURE 4.9 The Resistor Component Properties Editor Window (from an Expert Mode).

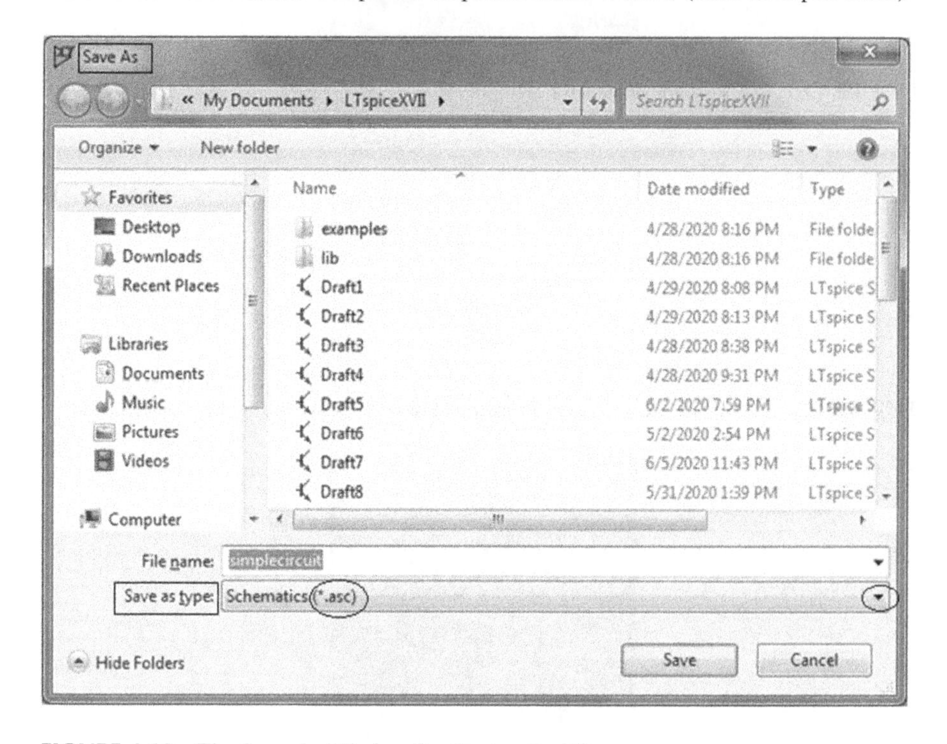

FIGURE 4.10 The Save As Window for Storing the Files.

The saved circuit schematic (.asc) named simplecircuit is shown in Fig. 4.11.

The LTspice XVII circuit schematic can be pasted into a word processor quite easily. The circuit when copied and pasted into an MS-Word document appears as shown in Fig. 4.12.

(**Note:** We can cut-and-paste the circuit by pressing Ctrl + C when the schematic window is active. The action copies the circuit schematic to the windows clipboard that can be pasted into a word processor document. Also, the entire LTspice XVII schematic window can be copied to the clipboard by pressing Alt + Prnt Scrn when the window is active.)

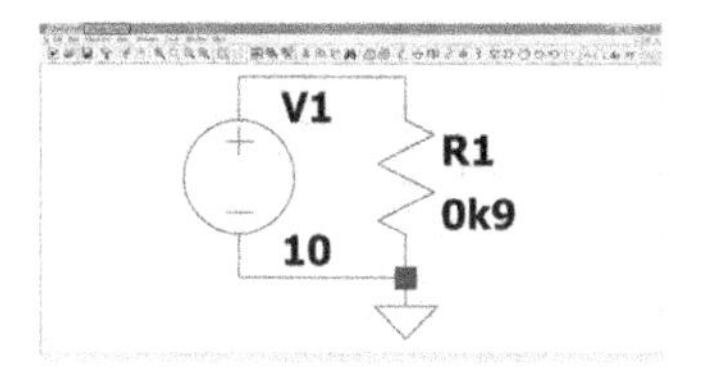

FIGURE 4.11 The Completely Drafted Circuit (Ready for Simulation).

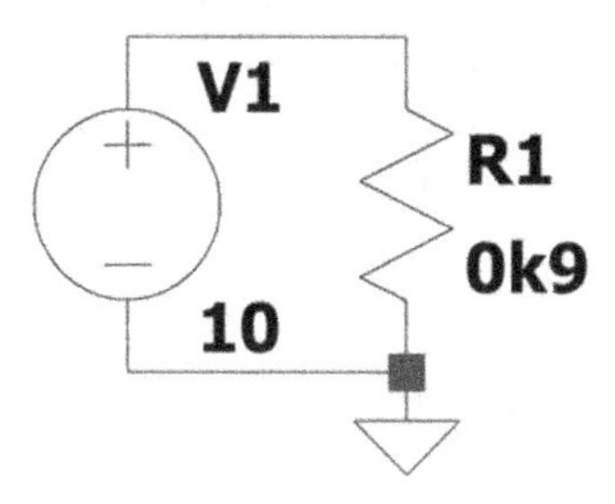

FIGURE 4.12 The Circuit Diagram in MS-Word.

The circuit is now complete. Before doing a simulation analysis, the user must ensure the following points:

- The circuit diagram is properly drawn and saved.
- There must not be any floating parts on the schematic diagram page (i.e., unattached devices).
- The components/device models must be assigned appropriate numeric values.
- There should be no extra wire.
- A common ground connection must be there in the circuit diagram.
- **Setting an operating point DC analysis for running a simulation as follows:**

v. After drafting the desired schematic circuit diagram in a complete manner, set up or configure simulation parameters for doing a basic DC operating point analysis.

From the LTspice XVII schematic menu, select the Simulate option, and then the Edit Simulation Cmd sub-menu option

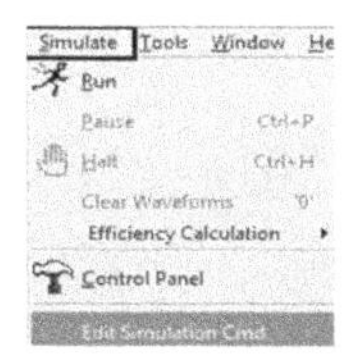

to open the Edit Simulation Command window dialog box. In the Edit Simulation Command dialog window, choose the tab DC op pnt from the tabs providing different simulation options for analyzing a circuit. Now, the SPICE directive .op appears in a text box at the bottom of the dialog window (Fig. 4.13).

After selecting the **DC op pnt** simulation tab option, click the **OK** (no need to set any parameter in the analysis). The SPICE directive statement .op (the dot command) gets attached to a mouse pointer or cursor (so that it can be dragged around and dropped with a click at someplace (preferably at the bottom) on the drafted schematic (Fig. 4.14).

vi. Click on the **Run** schematic button (a little running man) to display the **.op** simulation results on the basic circuit diagram. In a short time, a separate text window (the Operating Point window) with a list of voltages at different nodes and currents flowing through different elements in a tabular form appears as shown in Fig. 4.15.

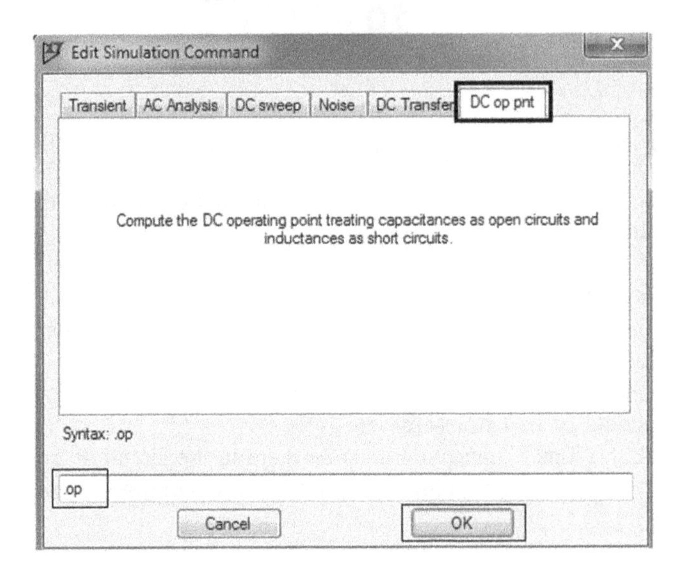

FIGURE 4.13 The Edit Simulation Command Dialog Window.

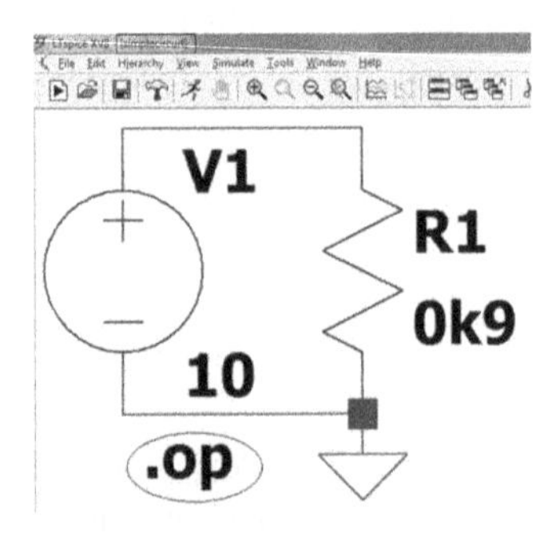

FIGURE 4.14 The .op Analysis for Simulating the Circuit Schematic.

- **Running the Simulation as follows:**

(**Note:** LTspice XVII automatically provides number labels for all the circuit connections or nodes such as N001, N002, N003, etc. The name of any node can be recognized by moving a cursor over a wire connected to the node of interest. The node name gets displayed on the left side of the Information Bar at the very bottom of the LTspice XVII schematic window. Because the default node numbers appear unfamiliar, it is better to assign names to the nodes before running a simulation. The ground node also has a label 0 associated with it.)

LTspice XVII uses nodal analysis to calculate voltages at all the nodes in the circuit relative to the reference ground node. If a ground node is not included in the circuit schematic, an error is received while performing a simulation.

- **The .op window text on a document**

The user can select a text in the Operating Point window by simply employing a window method of scrolling the mouse over all the text to be copied onto a clipboard while holding a mouse button and pressing Ctrl + C. Close the output window (optional) and paste the copied text into any MS-word document as shown as follows:

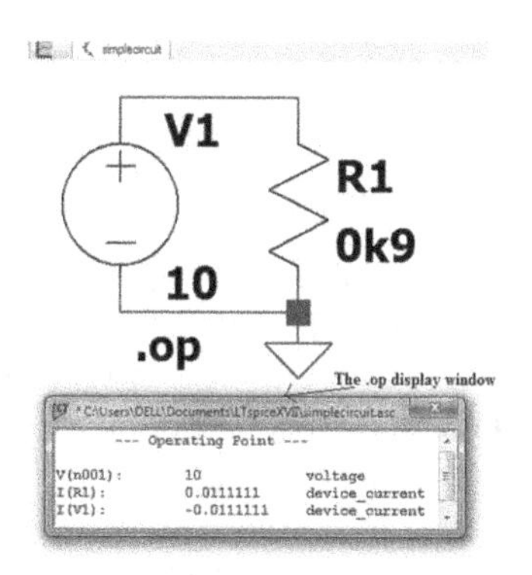

FIGURE 4.15 The DC Numeric Voltages at the Nodes and Currents through the Elements.

```
V(n001):          10               voltage
I(R1):            0.0111111        device_current
I(V1):            -0.0111111       device_current
```

The first line, V(n001): 10 voltage, states that a parameter is a voltage and a voltage at the node N001 (generic node 1) with reference to the ground node, that is, voltage across the R1 is 10 V (here, V is the default unit).

(**Note:** If the user does not explicitly label the circuit nodes, LTspice XVII automatically labels the nodes with specific numbers.)

The second line, I(R1): 0.0111111 device_current, states that a parameter is current and a current flowing through the component R1 is 0.0111111 A (or 11.1111 mA).

The current flowing from the voltage supply is –0.0111111 (or −11.1111 mA), where a minus sign indicates that the current is flowing out of the voltage source positive terminal.

The given results verify an Ohm's law, that is, V1 = I(R1) x R1 as follows:

$$10V = (0.0111111) \times (0.9 \times 10^3) = 9.99999V$$

Here, the current through the V1 is negative because LTspice XVII follows a passive sign convention and assumes a current to be negative if entering into a negative terminal of a device and leaving out from its positive terminal.

4.2.2 Bias Point Voltage Labels for Nodes

The simulator assigns a node number to each connection in the drafted circuit diagram. But, at some places, it is useful to display a numeric DC (voltage/current) value of the node.

To display a DC voltage value of the connection point as the circuit node label/name, place the corresponding operating point data value as the net/node label on the schematic. Thus, after running a simulation, if it is required to label/name the drafted circuit's nodes with voltage, current and power values at the connection points, do as follows:

Operating Point Data Labels: To display the operating point voltage of a generic node N001 after closing the Operating Point window simulation results, first move a cursor near the node (Fig. 4.16) and left-click on it when a cross-pointing cursor changes into a **red-colored** voltage probe cursor.

The operating point DC voltage value, that is, 10 V gets displayed as the net label as shown in Fig. 4.17.

Thus, a voltage across the R1 with reference to a common ground (the global node 0) is 10 V.

Alternatively, to label the node with the operating point DC value, after running the .op simulation, double-click the desired wire segment so that the operating point voltage of the node gets displayed.

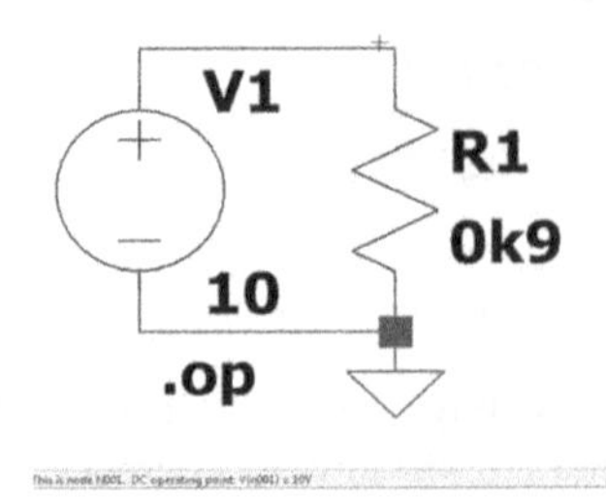

FIGURE 4.16 Probing a Voltage at the Selected Node for Labeling.

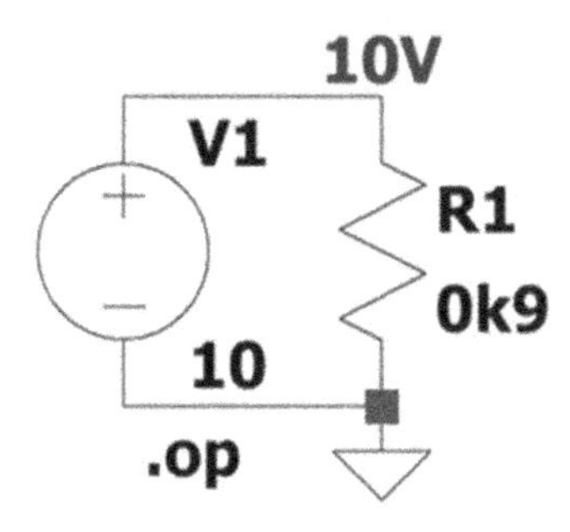

FIGURE 4.17 The Operating Point Voltage as the Node Label.

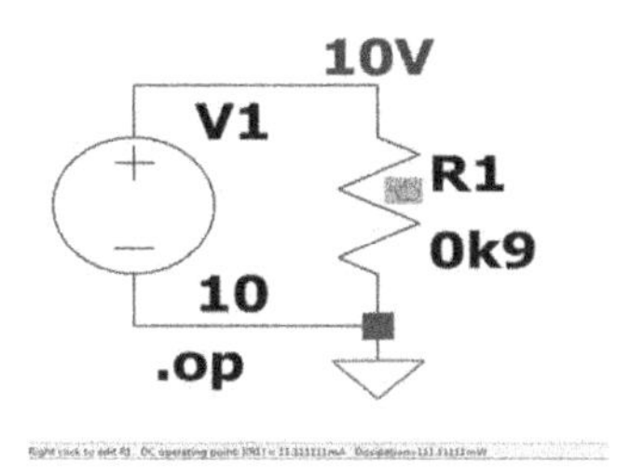

FIGURE 4.18 Reading the Current and Power Dissipated in the Information Bar.

(**Note:** Right-click on a data value label to change a number of significant figures in the displayed operating point numeric value, and type as round(trace_label*100)/100 in the **Edit expression to display** text box to for having an answer up to two decimal places. For retaining three significant figures, multiply and divide by 1000, and so on.)

Now, bring a cursor near to the R1 until a cross-hair pointing cursor changes into a pointing finger to display current flowing through it along with power dissipated in the Information Bar at the bottom of the schematic window (Fig. 4.18).

Observing a message in the Information Bar (Fig. 4.18), it can be said that a current flowing through the R1 is 11.111111 mA and power dissipation is 111. 11111 mW.

4.2.3 Nodes Labeling using Current (or Any Expression) Values

First, right-click over the operating point numeric net label, that is, **10V** displaying the voltage at a generic node N001. The following **Displayed Data** window appears on a screen:

The target is to use an operating point current value flowing through a generic node N001 (or displaying the computed value of an expression entered) as the net label for it. In Fig. 4.19, the default expression in the text box is a dollar character $ (displaying a voltage value as the label for that node).

Now, to display a current through the R1 as the node label, delete the $ and click on the I(R1), and then click on the **Evaluate, Copy to Clipboard, and Quit** (or the **OK**) as shown in Fig. 4.20.

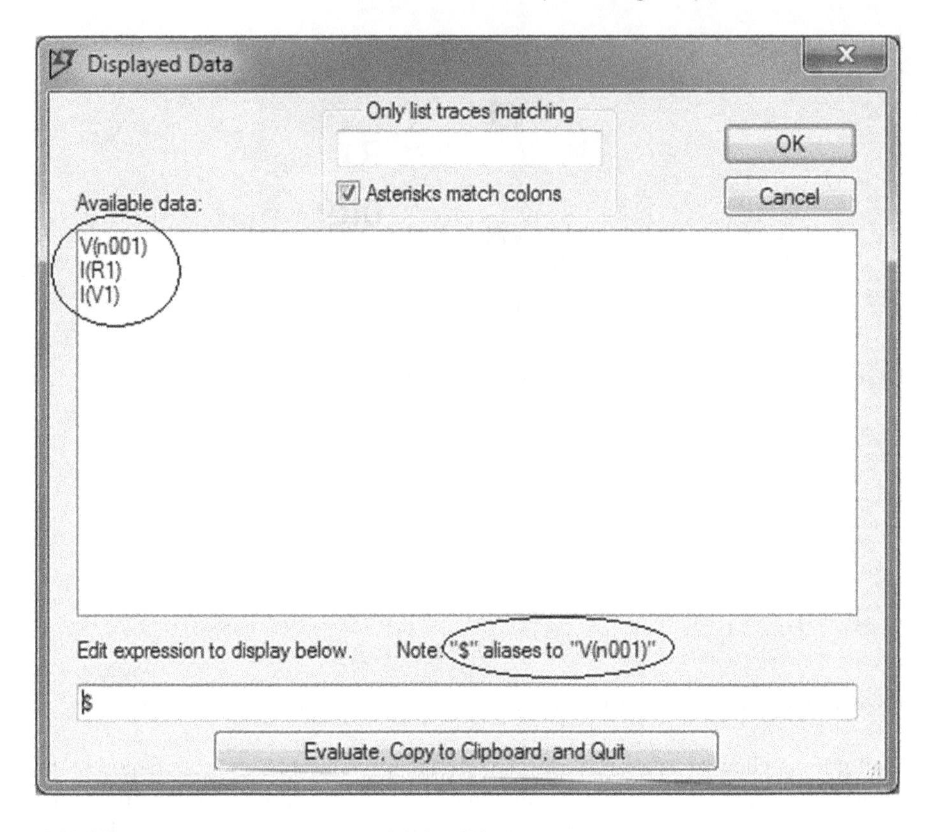

FIGURE 4.19 The Displayed Data Dialog Window.

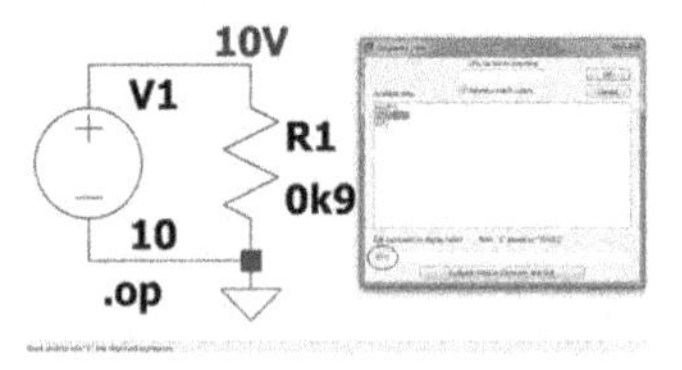

FIGURE 4.20 Using the Available Data Expression to Display a Current as the Net Label.

The value of a current through the R1, that is, 11.111111 mA gets displayed as the node label as shown in Fig. 4.21.

(**Note:** The numeric value data labels can be moved anywhere by using the Move schematic tool.)

It is to be noted that in Fig. 4.19, any expression (or equation) can also be entered in the text box to display the power dissipated through the R1, that is, V(n001)*I(R1) or voltage between two points or nodes, that is, V(n001)-V(n002) as the labels for nodes. Alternatively, the user can also use comma-separated node names, that is, V (n001,n002) to display a voltage between two points as the net label.

- **Labeling using both the voltage and current values:** Now, if the user double-click the vertical wire segment (or a small length wire segment

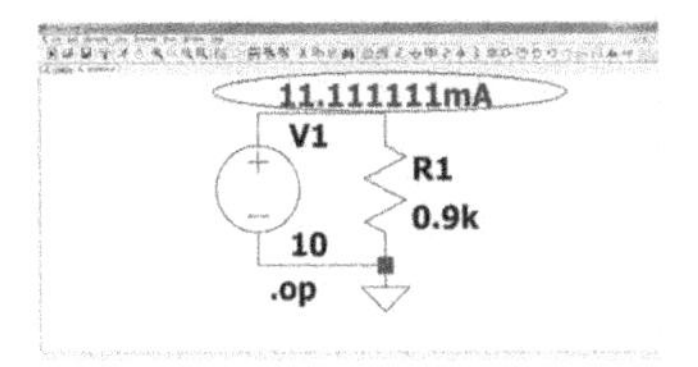

FIGURE 4.21 The Operating Point Current as the Node Label.

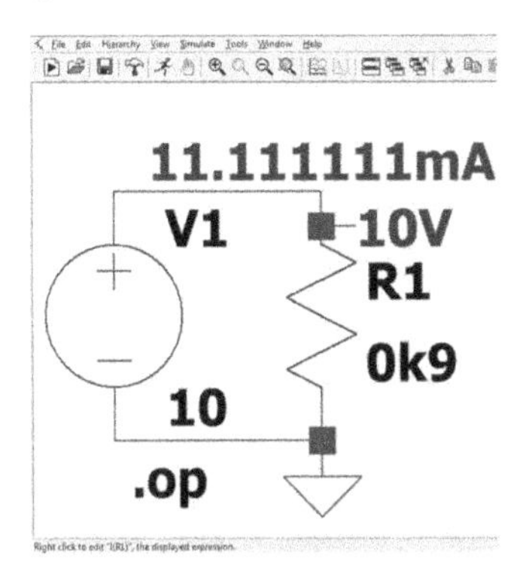

FIGURE 4.22 Labeling the Node with the Operating Point Voltage and Current Values.

connected to it for better looks), both the voltage and current values get displayed as the labels for a generic node N001 as shown in Fig. 4.22.

As the current and power values are not integers, it is always better to limit the bias point values in non-integer form to some specified decimal places using a rounding technique.

- **Using the round function:** It is sometimes useful to apply the round function to truncate a number of digits displayed, for example, to limit the current through the R1 to 3 decimal places, type the expression round(I(R1)*1k)/1k, whereas for limiting up to two decimal places, multiply and divide by 100)

Looking at the Operating Point window's list of data values, it can be stated that the current flowing through the R1 is 0.0111111 A as shown in the following:

I(R1): 0.0111111 device_current

Right-click over the current label saying **11.111111mA** to open the Displayed Data dialog window and type the round(I(R1)*1k)/1k in the available field for editing an expression as shown in Fig. 4.23.

The earlier-typed round function rounds off the 0.0111111 A up to three decimal places, that is, 0.011 A (or 11 mA), and then display the result as shown in Fig. 4.24.

On the other hand, if we type the **round(I(R1))** for no decimal places, the command rounds off the 0.0111111 A to zero decimal places, that is, 0 A (or 0 mA), and then display the result as shown in Fig. 4.25.

Thus, the round function may be used to truncate the number of digits displayed.

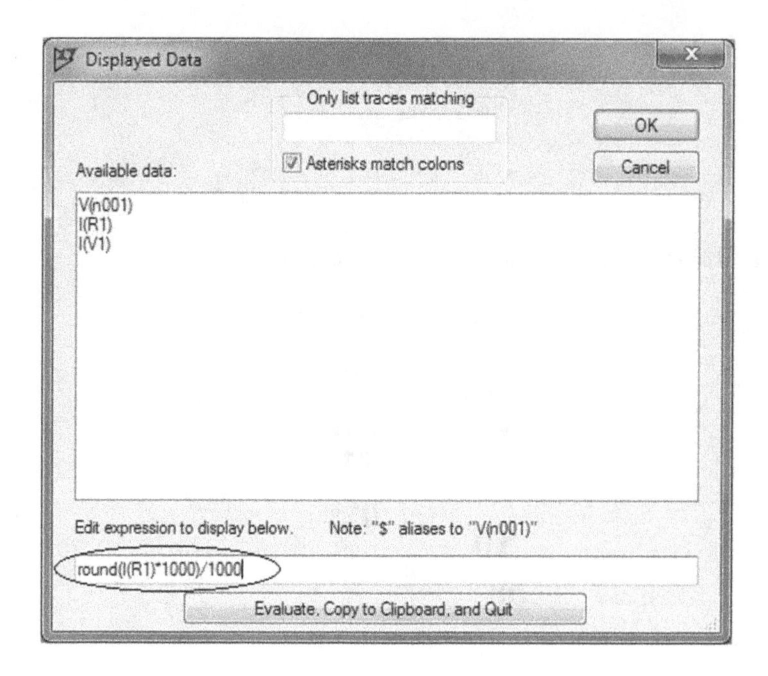

FIGURE 4.23 The Displayed Data Dialog Window.

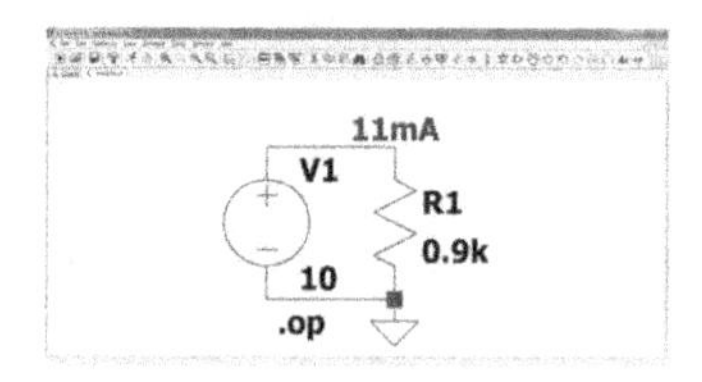

FIGURE 4.24 The Rounded Current Label.

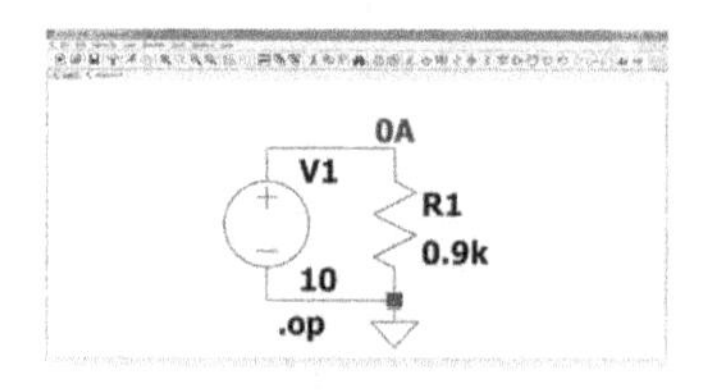

FIGURE 4.25 Rounding the Numeric Current Label to No (zero) Decimal Places.

4.2.4 User-Defined Expressions and Parameters Evaluation

- **Expressions with Constants or User-Defined Variables or Functions**

To set or change values of the component at once, and thereby avoiding editing individual component values every time or to define parameters in terms of expressions having constants or other global parameters/functions of other global parameters, the .param directive is used. The parameters are then used to define the values of the components in their value fields.

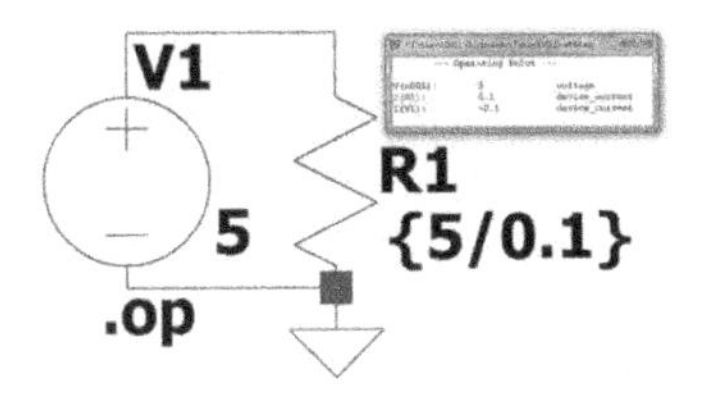

FIGURE 4.26 The Expression in Curly Braces for the Component Value.

When an expression enclosed in curly braces { } is encountered for the component value, the simulator evaluates the entered expression before a simulation begins based on all the relations available at the scope and reduces it to a floating-point value as shown in the following list:

- **Expressions having Constants for the Component Parameter Value**

An expression containing mathematical operations on constants using the valid operators for the component value can be calculated as shown in Fig. 4.26:

An expression containing a variable name along with the constants for the component value can also be calculated in real-time as shown in Fig. 4.27:

Here, the resistance parameter R of the resistor R1 is set equal to the node voltage divided by 2, where the node voltage is referred to as V(N001). Thus, set the value of the resistor R1 to R = V(N001)/2. Now run the .op simulation.

Without curly braces { }, the expression is not reduced to a value before simulation but is calculated during a simulation in real-time.

- **Expressions with Operations on User-Defined Variables**

The **.param** directive allows creating **user-defined** variables that are useful for varying the component values without actually editing the component properties (Fig. 4.28).

The .param statement tells the simulator that R is a global parameter.

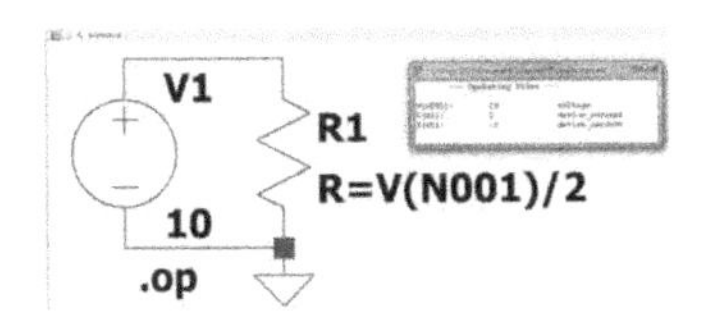

FIGURE 4.27 Setting the Resistance Parameter with the Voltage Source.

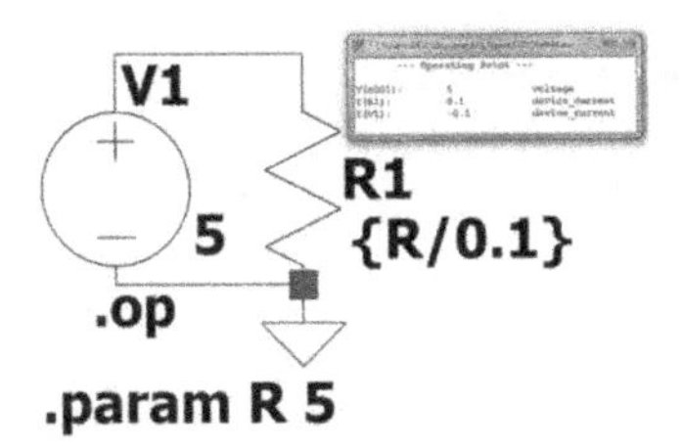

FIGURE 4.28 Evaluating the Expression Containing a Variable.

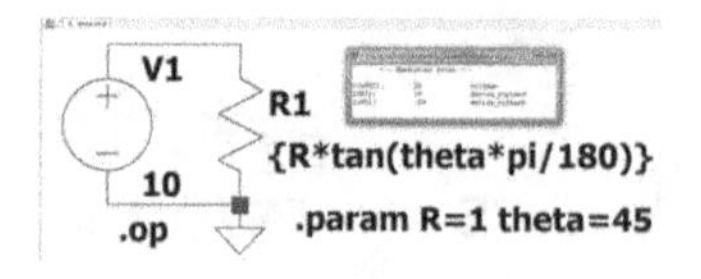

FIGURE 4.29 Evaluating the Expression Containing a Variable and the Built-In Functions.

- **Expressions with the Built-in Functions on User-Defined Variables**

In LTspice XVII, trigonometric functions assume an argument in radians and the pi/180 term should appear in the argument of the tan function to convert degrees into radians.

The .param statement can be used to define more than one variable and storing values in them as shown in the following discussion:

Place the blank .param directive on the schematic. Now, press a right mouse button on the placed directive to open its GUI for editing the parameters (see Fig. 4.30). Click the OK button to place the complete directive statement and press the Run schematic tool button to see the simulation results as shown in Fig. 4.29.

4.3 SIMULATING CIRCUIT HAVING CURRENT SOURCE

i. Creating the circuit diagram and labeling nodes: After constructing a resistive circuit with an independent current source and voltage source as shown in Fig. 4.32. To label all the drafted circuit nodes explicitly, select the **Label Net** tool (a character **A** in a box) with a left-click. This opens-up the **Net Name**

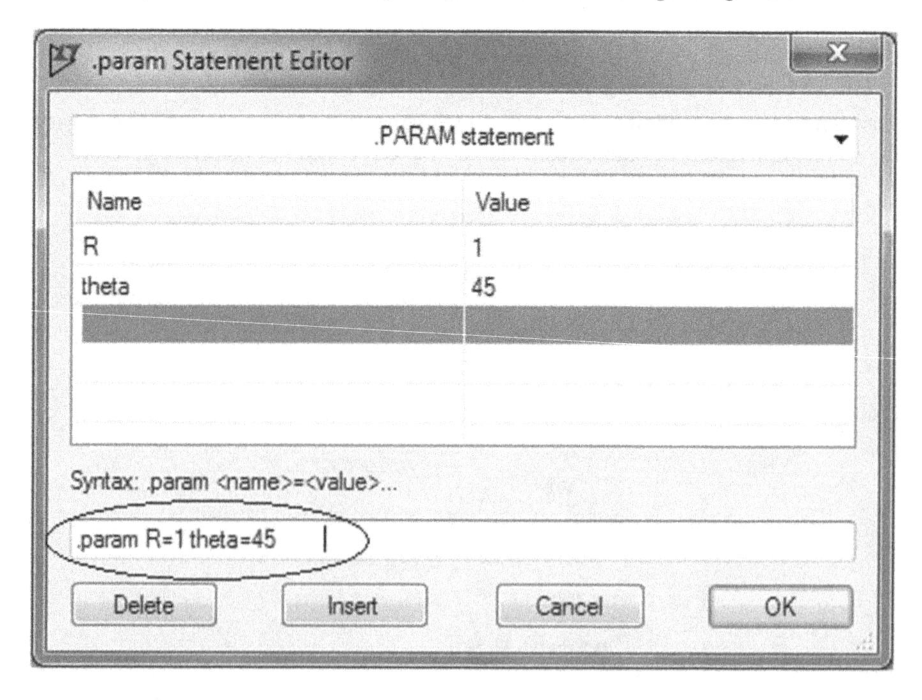

FIGURE 4.30 The .param Statement Editor Window.

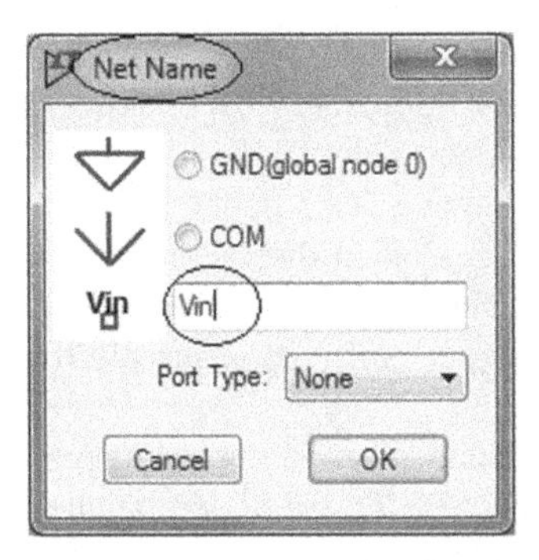

FIGURE 4.31 The Net Name Window for Labeling the Circuit Nodes.

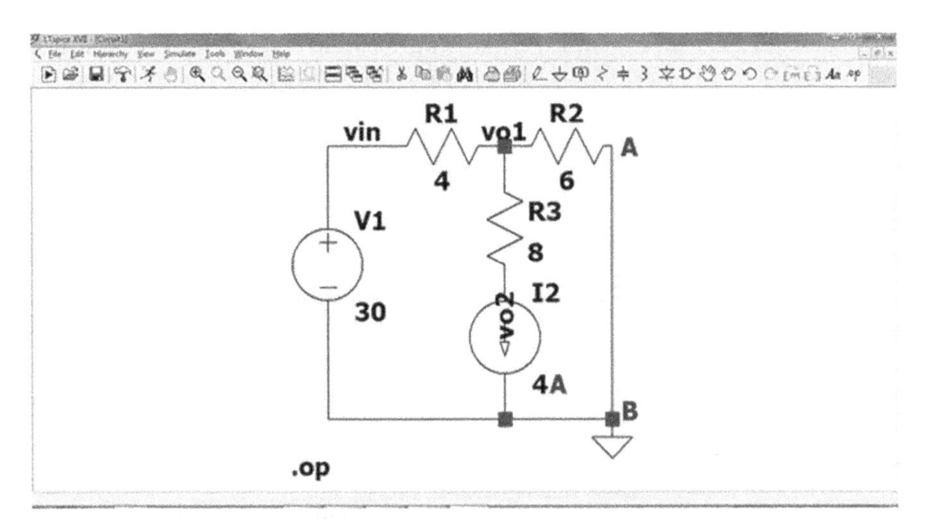

FIGURE 4.32 The Ready for Simulation Complete Circuit Schematic.

window (see Fig. 4.31) having a field box in which the user can type in an appropriate label for the circuit nodes, and afterward click the OK.

The net label gets attached to a cursor and can be placed at any point on the schematic and wires using a catching point (the blue colored little rectangle on the center of the lower edge e.g., Vin) Place (by clicking) the little rectangle on the specific node or wire and it gets automatically connected. This helps the user to easily identify the node voltages.

(**Note:** Spaces are not allowed while labeling node names.)

i. After completing labeling the circuit nodes, again select the simplest mode known as a DC operating point to place the .op SPICE directive on the schematic as shown in Fig. 4.32.

Remember: It is always preferable to label the circuit nodes (using the Label Net tool) to make it easier for the user to find them and identify which node is representing a specific terminal of which device in the simulation results. Otherwise, the SPICE netlist (generated by clicking the View -> SPICE Netlist options from the schematic menu) allows the user to see the node numbers (default or assigned ones) for each device or component along with the component value.

The circuit diagram can be copied directly to MS-Word by selecting the **Copy bitmap to Clipboard** sub-menu from the Tools schematic menu. The file can also be saved as an image into a dot emf (i.e., .emf) file for sharing purposes.

(**Note:** LTspice XVII automatically creates an appropriate SPICE directive depending on the inputs. The SPICE directives are the instructions applied to the simulation program itself (not the parts of the circuit) to direct the simulator on what to do with the circuit using the directive language.)

ii. Now run the simulation using the **Run** menu by clicking on the icon that looks like a runner. A new window (the Operating Point text window) opens up to display the results of the simulation as shown in Fig. 4.33.

The first line, V(vin): 30 voltage, states that the value of a parameter voltage at the node vin (generic node 1 with the default reference number N001) is 30 volts. Similarly, a voltage at the node vo1 is 8.4 volts.

Also, a voltage at the node vo2, that is, the voltage across the current source is –23.6 volts, where a minus sign indicates that an arrow of the current source points in a downward direction.

The fifth line, I(R3): 4 device_current, states that the value of a parameter current through the resistor R3 is 4 amperes (the results are displayed in the default units).

The current supplied by the voltage supply is –5.4 A, where a minus sign indicates that a current is flowing out of the voltage source.

(**Note:** The electrically common nodes (or common connection points) in the circuit diagram always share common reference numbers by default. The DC node voltages represent the readings of the parameter voltage from the respective node

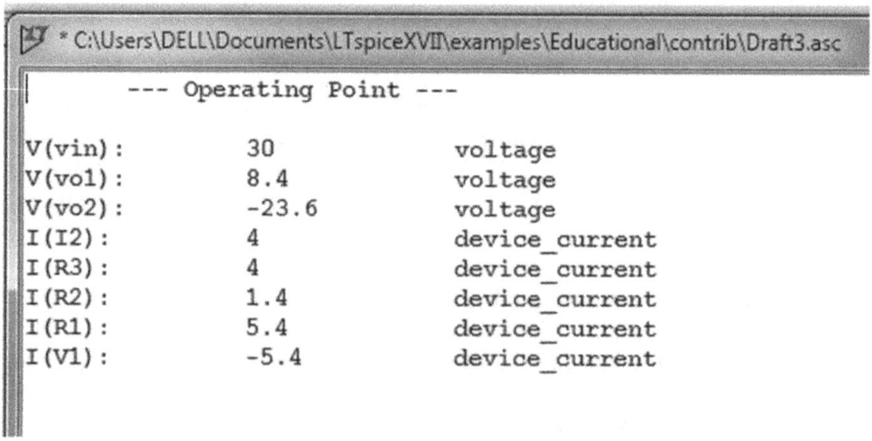

```
* C:\Users\DELL\Documents\LTspiceXVII\examples\Educational\contrib\Draft3.asc

       --- Operating Point ---

V(vin):          30            voltage
V(vo1):          8.4           voltage
V(vo2):          -23.6         voltage
I(I2):           4             device_current
I(R3):           4             device_current
I(R2):           1.4           device_current
I(R1):           5.4           device_current
I(V1):           -5.4          device_current
```

FIGURE 4.33 The Operating Point Result Window.

(having a specific reference number or an explicitly assigned name) to the common ground node having a reference number 0.)

4.4 VOLTAGE DIVIDER SIMULATION (BIAS POINT AND DC SWEEP)

4.4.1 Using DC Operating Point Analysis

i. Construct the circuit diagram having a voltage source connected in parallel to the two resistances in series and save it as voltagedivider.asc. Below is the completely drafted (ready for simulation) circuit with assigned values to the parameters of the components and the net labeling:
ii. The results of the circuit simulation using a DC operating point analysis are shown in Fig. 4.34.

- **How to copy the .op text window to a clipboard:**

iii. The user can copy a list of the operating point simulation results in a tabular form to a clipboard by pressing Alt + Prnt Scrn keys while the given display window (the .op result window and the schematic) is active and paste it into an MS-Word document as shown in Fig. 4.35:

(**Note:** In the given list, the symbol V with an argument enclosed in parenthesis is used to denote a voltage parameter at a point, where an argument is a specific generic or any user-defined node number.)

Thus, the V(n001) denotes a voltage at a generic node N001 with reference to the common ground node 0 and is equal to 20 volts. The voltage across the R1 is V(n001) –V(out) = 20 – 13.3333 = 6.667 volts.

Remember: To display a voltage between the two points or nodes, say N001 and N002, the user can use comma-separated generic node numbers, that is, V(n001,n002) or type an expression V(n001)–V(n002).

As per a voltage divider rule, the voltage across the R1 should be equal to: $20 \times 500/(500 + 1000) = 6.66666666667$ V.

Thus, we can say that the simulation results are in agreement with a voltage divider rule for two resistances.

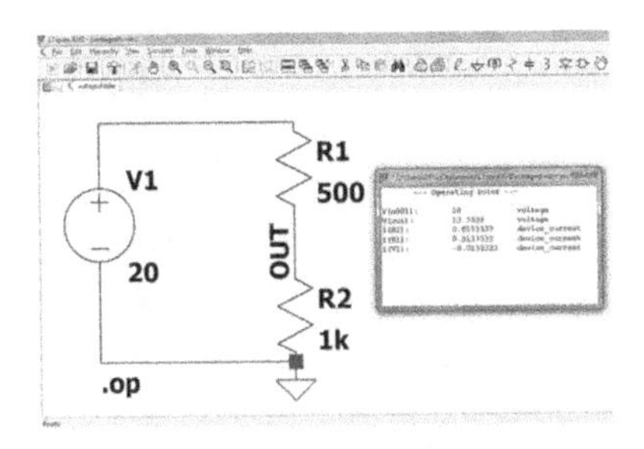

FIGURE 4.34 The .op Simulation Results of the Voltage Divider Circuit Schematic.

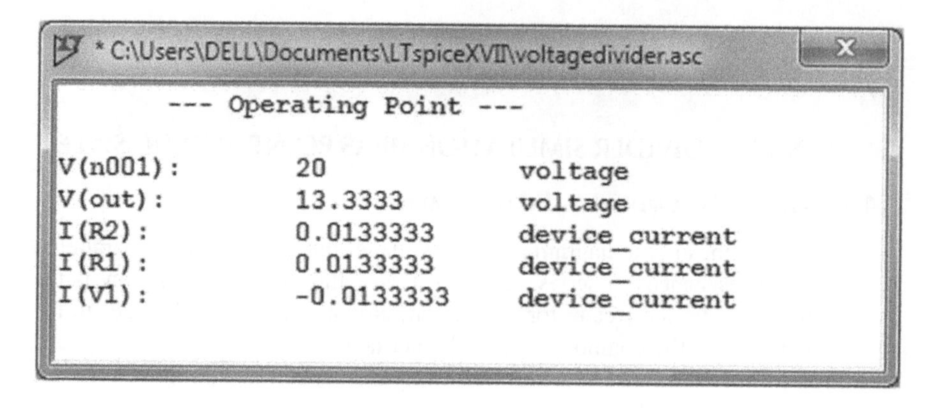

* C:\Users\DELL\Documents\LTspiceXVII\voltagedivider.asc
--- Operating Point ---
V(n001): 20 voltage
V(out): 13.3333 voltage
I(R2): 0.0133333 device_current
I(R1): 0.0133333 device_current
I(V1): -0.0133333 device_current

FIGURE 4.35 The .op Text Window in a Document.

iv. Close the .op result window. Now, bring a cursor on the node OUT to observe a DC voltage across the R2 in the Information Bar as shown in Fig. 4.36.

- **Reading Current and Power Values in the Information Bar**

v. Close the .op window and delete the node label OUT. Now, moving a cursor on the R2 (component body) provides the current and power dissipation information at the Information Bar when a pointing finger appears on the R2 as shown in Fig. 4.37.

- **Using a Numeric Voltage for the Node Labeling:**

vi. After closing the .op window (appears after running the simulation) and deleting the node label OUT using the **Cut** schematic tool, the following output appears if the user clicks on the output node or the vertical wire segment (to which both the resistance terminals are connected) (Fig. 4.38).

- **Rounding the Operating Point Data Label:**

vii. Right-click over the displayed voltage label value saying 13.3333V (V is the default unit of voltage) to open the Displayed Data dialog window and type

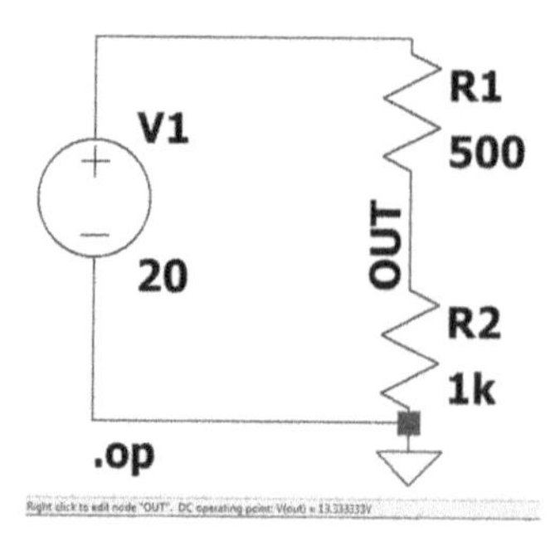

FIGURE 4.36 Reading the Node Voltage Value in the Information Bar.

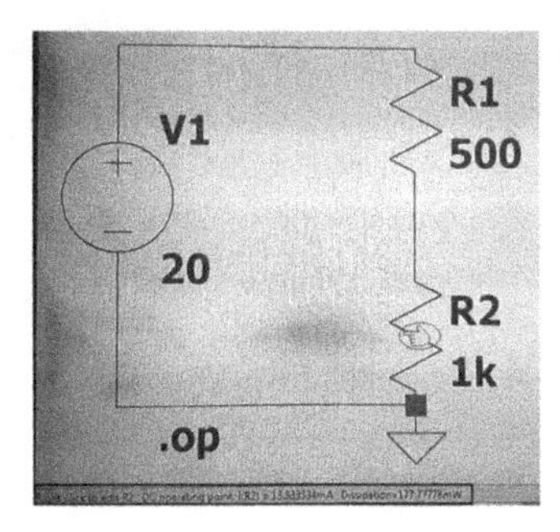

FIGURE 4.37 Reading the R2 Current and Power Dissipation Values.

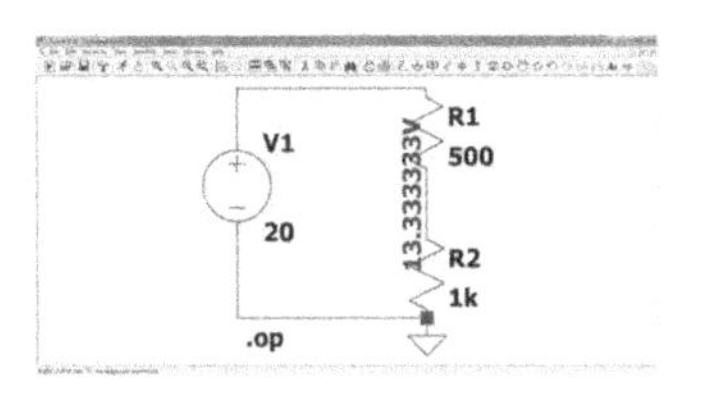

FIGURE 4.38 Voltage Value Displayed as the Net Label.

round($*1000)/1000 in the available text box under the option for editing an expression as shown in Fig. 4.39.

The earlier-typed round function rounds off 13.3333 V (as stated in the .op result window) up to three decimal places, that is, 13.333 V, and then displays the desired result.

viii. The non-integer current value as the node label can also be rounded-off by replacing a dollar character and typing round(I(R2)*1k)/1k command in the **Edit expression to display below** text box. The entered command rounds off 0.0133333 A DC current value (as stated in the .op result window) up to three decimal places, that is, 0.013 A (13 mA), and then displays the result as shown in Fig. 4.40.

To conclude, when the displayed data label values are in their default units, the round function gives the intended results.

4.4.2 Voltage Divider with Variable Resistor

A circuit can be examined in LTspice XVII either by changing the value for a particular parameter manually and then re-simulate the circuit to view a response or by using the .step command to sweep across a range of values in a single simulation run. Thus, the .step command can be used with the .op command to simulate the circuit with variable resistance.

i. In the earlier-drafted circuit saved with a file name voltagedivider, replace numeric values for the resistance parameter values of the R1 and R2 with

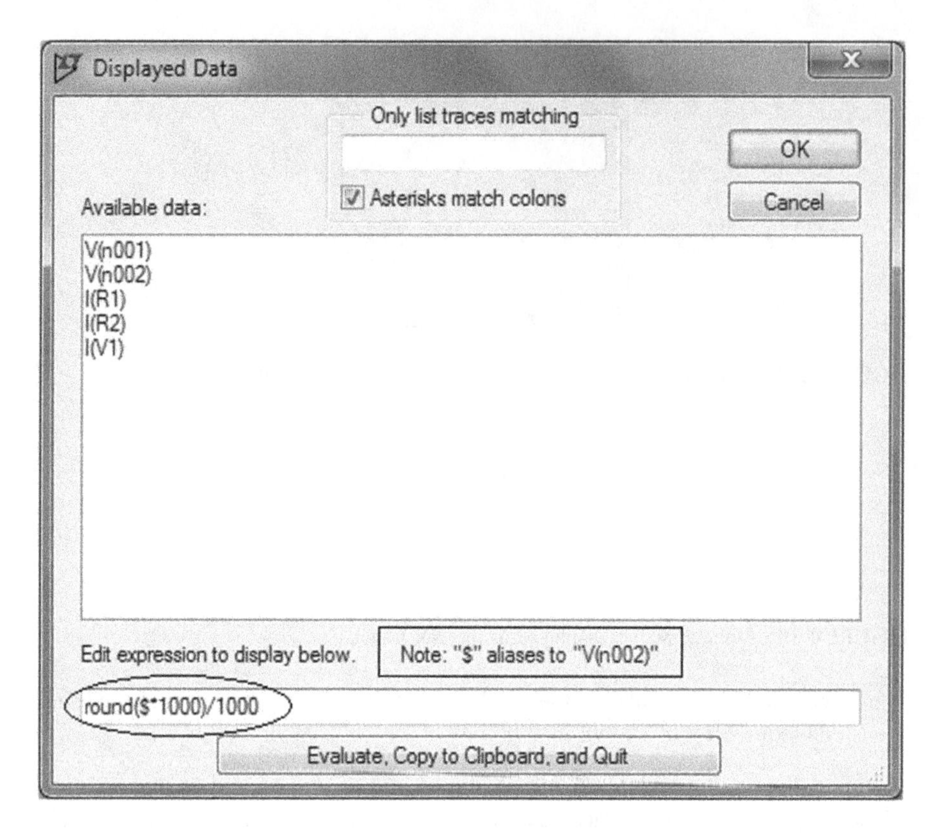

FIGURE 4.39 Rounding the Voltage Value Displayed as the Net Label.

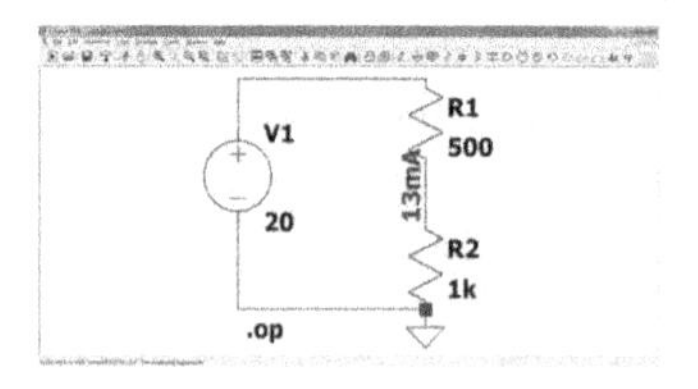

FIGURE 4.40 The Rounded Current Value as the Net Label.

variable names 10k-R and R encased in curly braces, respectively. Save the file with another name, say varresdivider. The action saves the changes made in a separate schematic file. Place the .step param blank directive command on the schematic as shown in Fig. 4.41.

ii. Now, edit the .step param command by right-clicking as shown in Fig. 4.42.

Here, the variable R is changed between 1 kΩ and 10 kΩ in increments of 1 kΩ using the .step command, whereas the variable 10k-R is changed between 9 kΩ and 0 kΩ in decrements of 1 kΩ.

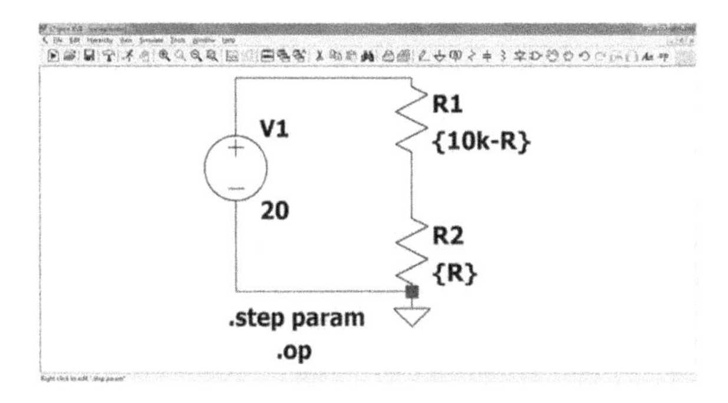

FIGURE 4.41 The Circuit Schematic Demonstrating a Variable Resistor.

.step Statement Editor

.step is used to overlay simulation results while sweeping user-defined parameters.

Name of parameter to sweep: R

Nature of sweep: Linear

Start value: 1k

Stop value: 10k

Increment: 1k

Syntax: .step param <Name> <Start Value> <Stop Value> <Increment>

.step param R 1k 10k 1k

Cancel OK

FIGURE 4.42 The .step Statement Editor Window.

Click the OK to place the complete .step param directive statement on the schematic.

iii. Run the .op (a DC operating point analysis) simulation for the circuit with the changed settings. A blank waveform viewer window appears as shown in Fig. 4.43.
iv. Now, probe the node voltages using the voltage probe cursor to observe the multiple .op simulation results for variable resistance values are shown in Fig. 4.44.

The plotted trace horizontal axis is the changed resistance value and the vertical axis is used to represent voltage values across the R1 and R2 divided from the variable resistors.

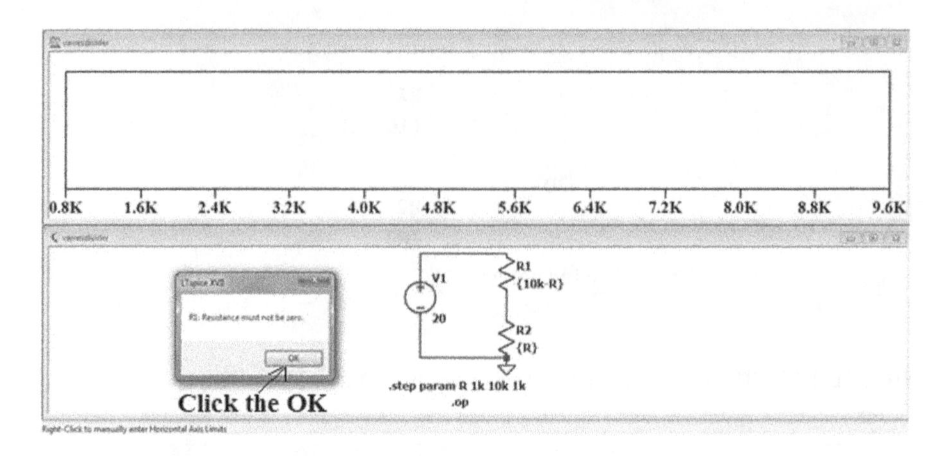

FIGURE 4.43 Stepping the Resistance Value with the .op Analysis.

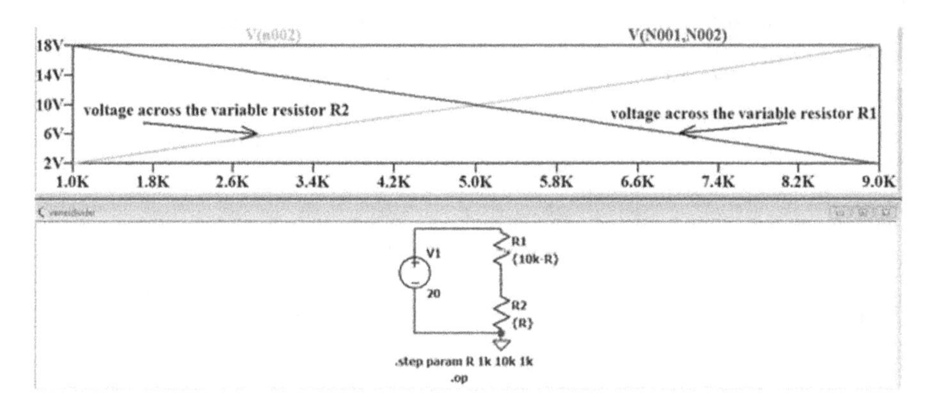

FIGURE 4.44 Node Voltage Variations against a Variable Resistance.

Thus, LTspice XVII allows stepping more than one variable at a time. But, only up to 3 parameters can be stepped simultaneously.

(**Note:** LTspice XVII includes an integrated waveform viewer to allow complete control on the way simulation data is used for plotting.)

4.4.3 DC Sweep Analysis of Voltage Divider

A DC sweep analysis (more advanced simulation analysis) allows the user to simulate a circuit many times for performing an operating point analysis of the circuit over a range of DC source (voltage or current) values and calculating an operating bias point at each specified value.

The analysis helps in obtaining DC characteristics of circuits (taking capacitors as open circuits and inductors as closed circuits) while sweeping only independent voltage or current sources. At a time, the user can sweep a maximum of three sources linearly or logarithmically by noting down their names.

- Simulating a voltage divider by sweeping the independent voltage source:

A DC sweep is performed by sweeping DC sources' values within a user-defined range using start values, stop values, and incremental steps for the pre-determined DC range so that a bias point can be computed for each value of the sweep.

Steps Involved:

i. Open the already saved schematic file named voltagedivider (with an extension .asc) by clicking on the **Open** schematic tool icon shown in Fig. 4.45.

This action brings the following window. Scroll down to select the file name or type it in the box next to the **File name** to open the file voltagedivider.asc as shown in Fig. 4.46.

Click the Open button to open the schematic file voltagedivider as shown in Fig. 4.47.

Choosing an Analysis Type: As the target is to observe changes in a DC voltage across the R2 (i.e., at a generic node N002 with reference to the common ground) when the V1 is swept from 0 to 20 volts, we need to select a DC sweep analysis. To achieve the target and plot the desired results, do as follow:

ii. Select the Edit Simulation Cmd sub-menu from the Simulate schematic menu (see Fig. 4.47). From the Edit Simulation Command window, select the **DC sweep** tab to open the Edit Simulation Command dialog box (Fig. 4.48) to sweep or vary a voltage value (DC parameter) of the source V1.

Now, set the parameters for sweeping the value of the voltage source V1 from 0 volts to 20 volts in increments of 2 volts as follows:

To set the V1 value to range from a minimum of 0 to a maximum of 20V, type V1 in the **Name of 1st source to sweep** text box, select Linear in the **Type of sweep** drop-down menu, type 0 (the minimum value) in the **Start value** box, type 20 (the maximum value) in the **Stop value** box and type 2 (the step size) in the **Increment box** as shown in Fig. 4.48.

i. After typing in all the inputs, click the OK and place the .dc V1 0 20 2 directive (automatically created by LTspice XVII on a basis of the given inputs in the Edit Simulation Command dialog box) on the circuit schematic as shown in Fig. 4.49.

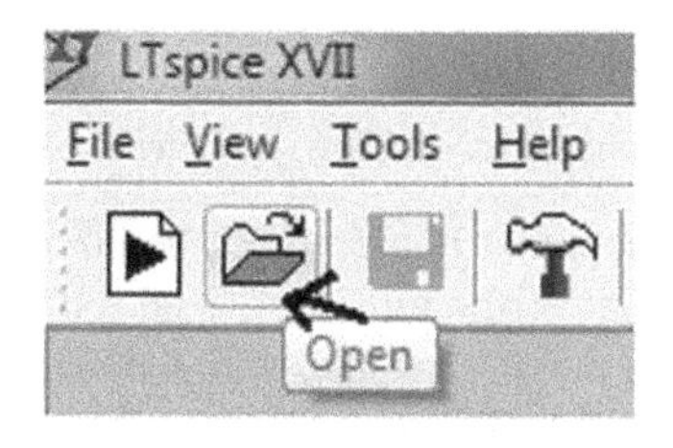

FIGURE 4.45 The Open Schematic Tool Icon.

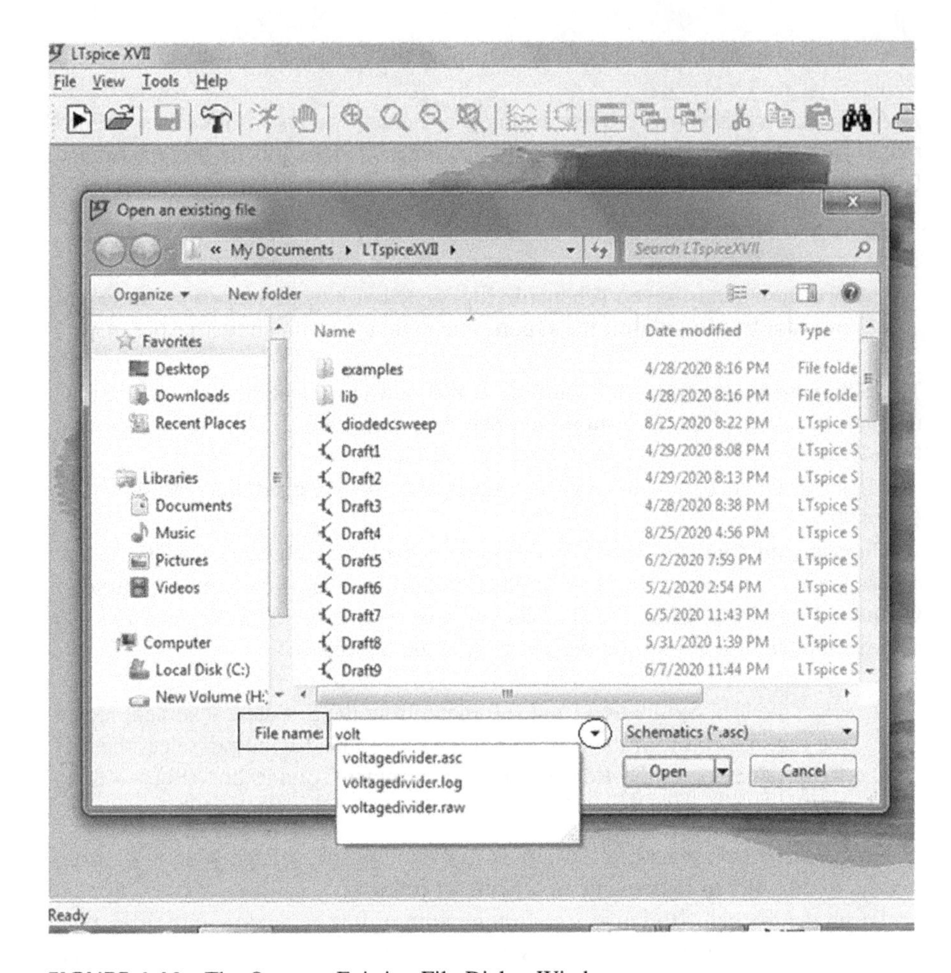

FIGURE 4.46 The Open an Existing File Dialog Window.

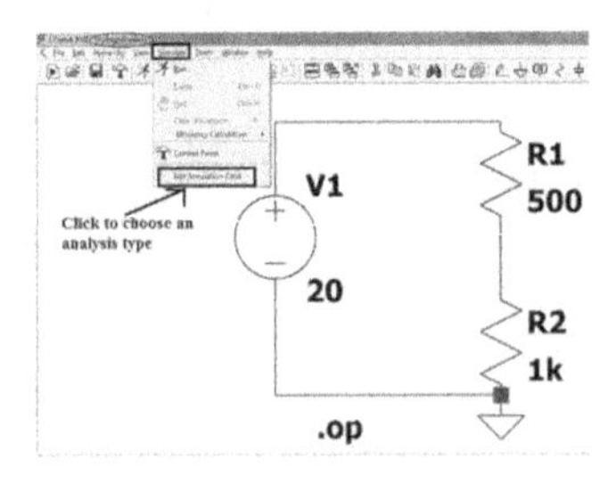

FIGURE 4.47 The Circuit Schematic with the .op Directive.

When the .dc analysis statement is placed on the circuit schematic, a semicolon automatically appears in front of the already placed .op statement instead of a dot (see Fig. 4.49). When a semicolon replaces a dot, an operating point analysis statement becomes inoperative and is treated just as a comment. Therefore, when a simulation is run the result of a DC sweep analysis gets displayed.

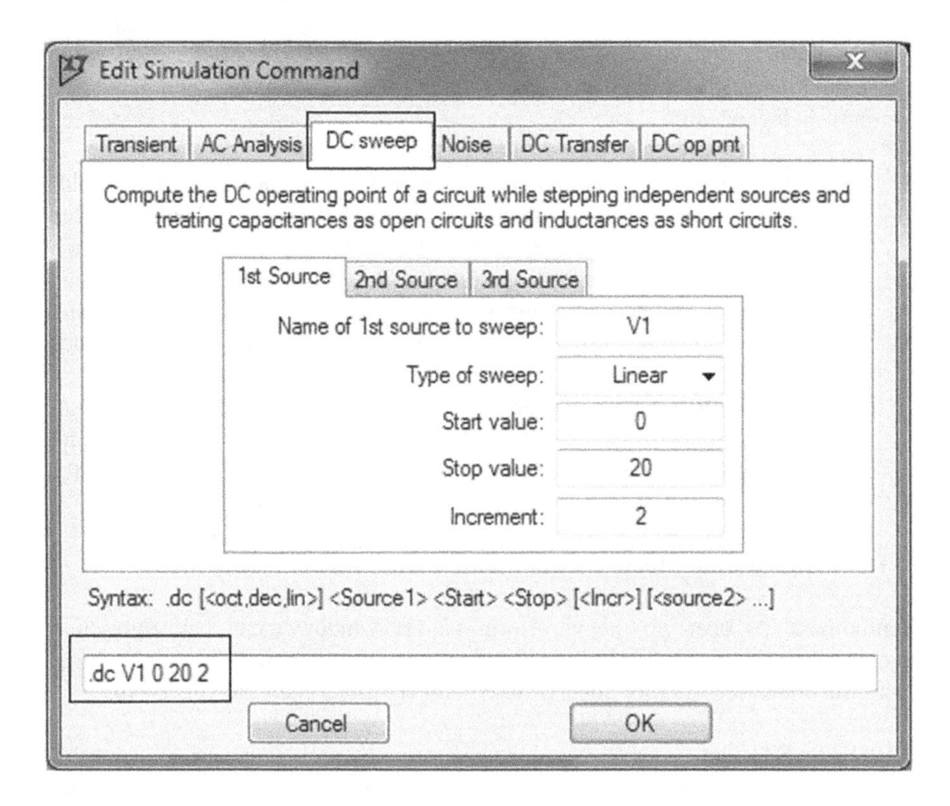

FIGURE 4.48 The Edit Simulation Command Dialog Box.

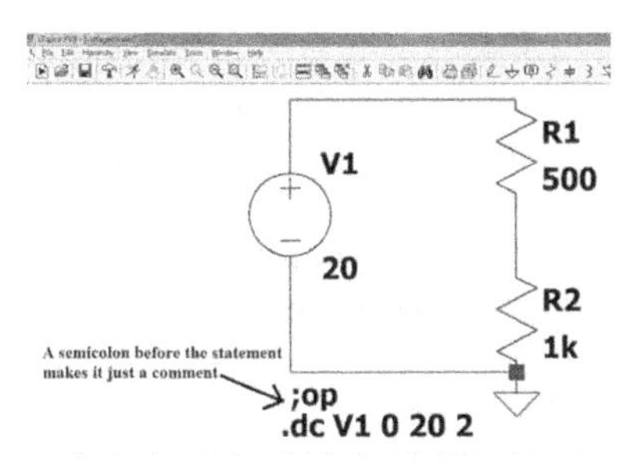

FIGURE 4.49 The Circuit using a DC Sweep Analysis.

(**Note:** When the .dc directive statement for doing a DC sweep analysis is placed on the schematic, LTspice XVII automatically places a semicolon before the previous .op (DC operating point analysis) directive statement by replacing a dot (or period) which indicates a termination of the .op analysis command.)

Remember: Before starting a fresh simulation, the previous simulation command (if not required, optional) can always be deleted using the **Cut** tool or selecting the **Edit -> Delete** options to remove the .op directive. The user can also terminate any directive statement manually by right-clicking on the analysis statement, say .op, and type a semicolon before the op command after deleting a dot.

Thus, the user can manually add and remove semicolons and periods to select the one simulation that is desired by right-clicking on each simulation command and editing it independently.

ii. Before running a simulation, change a waveform or trace background color to white by going to the schematic menu bar and clicking the **Tools** -> **Color Preferences** options. The **Color Palette Editor** dialog window appears. Now, under the **WaveForm** tab, select the **Background** from a scroll-down menu next to the **Selected Item** field, and then move the Red, Green, and Blue sliders to the extreme right (at a 255 RGB value) for changing a trace background to white.
iii. Click on the **Run** button to execute a given DC sweep simulation command. A blank graph window (the waveform viewer) with a white background appears above the schematic window. Now, probe the schematic circuit for plotting a desired electrical quantity in the waveform viewer.

- **Change the waveform viewer background color to white:**

Remember: To open up the waveform viewer window menu bar, right-click somewhere inside a plot pane. To plot the result with grid points on the plot pane, go to the View -> Grid plot context-based and right-tick the Grid check box.

IMPORTANT: Plot traces to examine waveforms by probing electrical quantities using the following actions in the schematic drawing along with a blank graph window:

- Left-click on a wire or node (a voltmeter probe appears) to plot node voltages.
- Left-click on the component/device model (a current probe having the shape of a plier generally used by electricians appears) to plot a current flowing through the component/device.
- Hold down an Alt key and left-click over the component/device model (a thermometer shape cursor appears) to plot a power dissipated in the component/device.
- Plot a voltage across the R2:

iv. To plot changes in a voltage referred to as V(n002) at a generic node N002 concerning changes in the V1, move a cursor near to the circuit node (i.e., N002). Left-click when a cross-hair pointing cursor changes into the red colored voltage probe cursor as shown in Fig. 4.50.

Now, release a mouse button to obtain a linear plot from 0 – 20 V on an x-axis and 0 – 14 V on a y-axis in the graph window as shown in Fig. 4.51.

Also, the .op directive statement can be deleted using the Cut schematic tool (optional). The .dc directive statement can be moved at a suitable position on the schematic using the Move schematic tool. To remove grid points, right-click anywhere inside the plot and go to the View to un-tick the Grid check box.

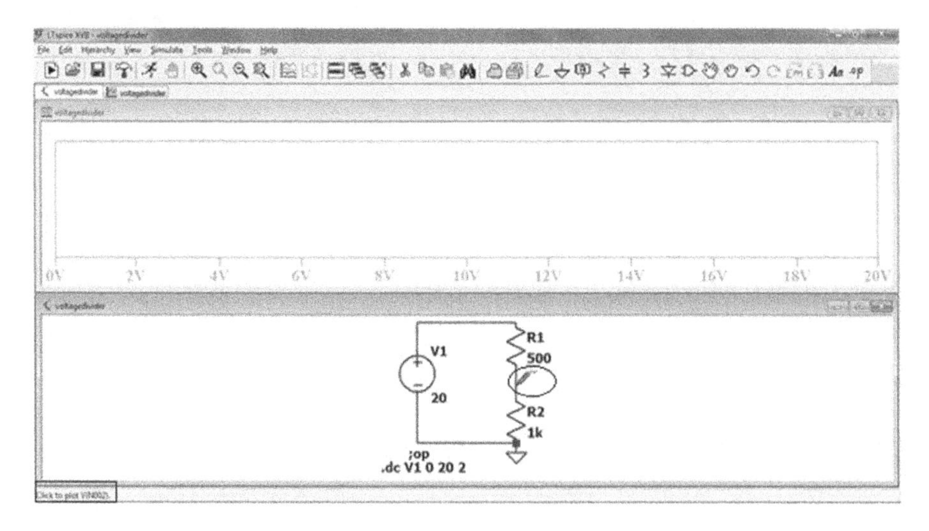

FIGURE 4.50 Simulating the Circuit using a DC Sweep Analysis.

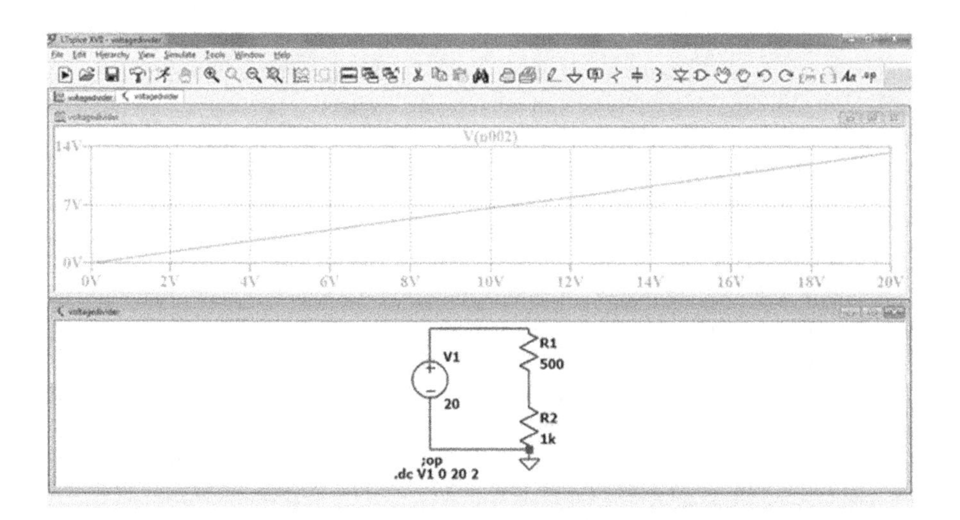

FIGURE 4.51 Plotting a Voltage across the Resistor R2 versus the V1.

According to a voltage divider rule for two resistors, a voltage drop across a resistor is directly proportional to the input voltage applied to the circuit, and therefore the graph is a straight line. A horizontal (x-axis) displays the voltage source V1, while a vertical (y-axis) is a plot of voltage at the junction of the R1 and R2 (or across the R2).

The user can right-click the waveform name, that is, V(n002) to have a drop-down box that allows attaching the 1st, 2nd, or both the numbered cursors as shown in Fig. 4.52. The cursors can be moved around with a mouse for reading individual

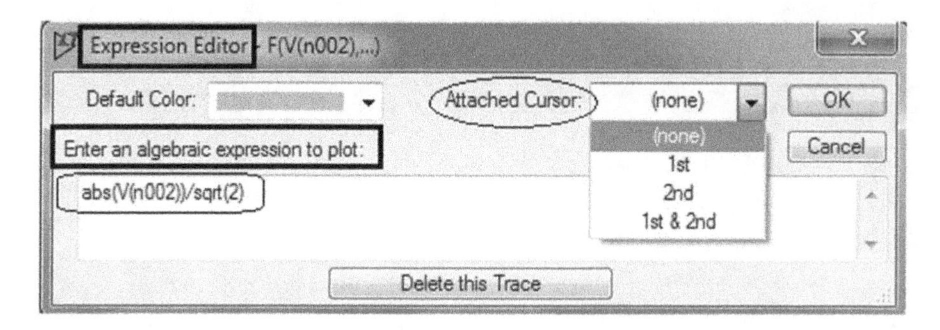

FIGURE 4.52 Using the Built-In Functions in the Expression Editor Window

values and their differences in time, frequency, and magnitude. The trace can also be deleted.

A modified expression can also be entered into the plotted trace, for example, the result of an absolute value of the voltage expression V(n002) divided by a square root of two can be plotted by typing **abs(V(n002))/sqrt(2)** in the Expression Editor text box as shown in Fig. 4.52, and then left-clicking the OK button.

While probing multiple traces, all curves can be plotted on the same pane (multiple scales on the same pane) or can be divided into separate panes for a better look or organizing information in an appropriate manner.

Plot a current trace:

v. To plot current flowing through the R2, that is, I(R2), first close the currently active voltage plot referred to by a name V(n002) and maximize the schematic window. Re-execute a simulation using the Run schematic button or tool to open a blank graphical window. Now, float a cursor on the body of the R2. When a cross-hair pointing cursor changes into the current probe cursor, left-click to plot changes in a current through the R2 concerning changes in the V1 in the graphical waveform viewer window. As a current through a resistor is proportional to a voltage across it, the output graph is a straight line where a horizontal x-axis displays the voltage source V1, while a vertical y-axis is a plot of current through the R2.

- **Edit a currently active plot for selecting other waveforms to plot:**

vi. After plotting a graph in the waveform viewer window, the user can simply replace the already plotted graph with a new trace by clicking on the **Pick Visible Traces** schematic icon (looks like a series of Cartesian graphs) and selecting the new trace (from the **Select Visible Waveforms** dialog box) to be displayed on the same waveform viewer window.

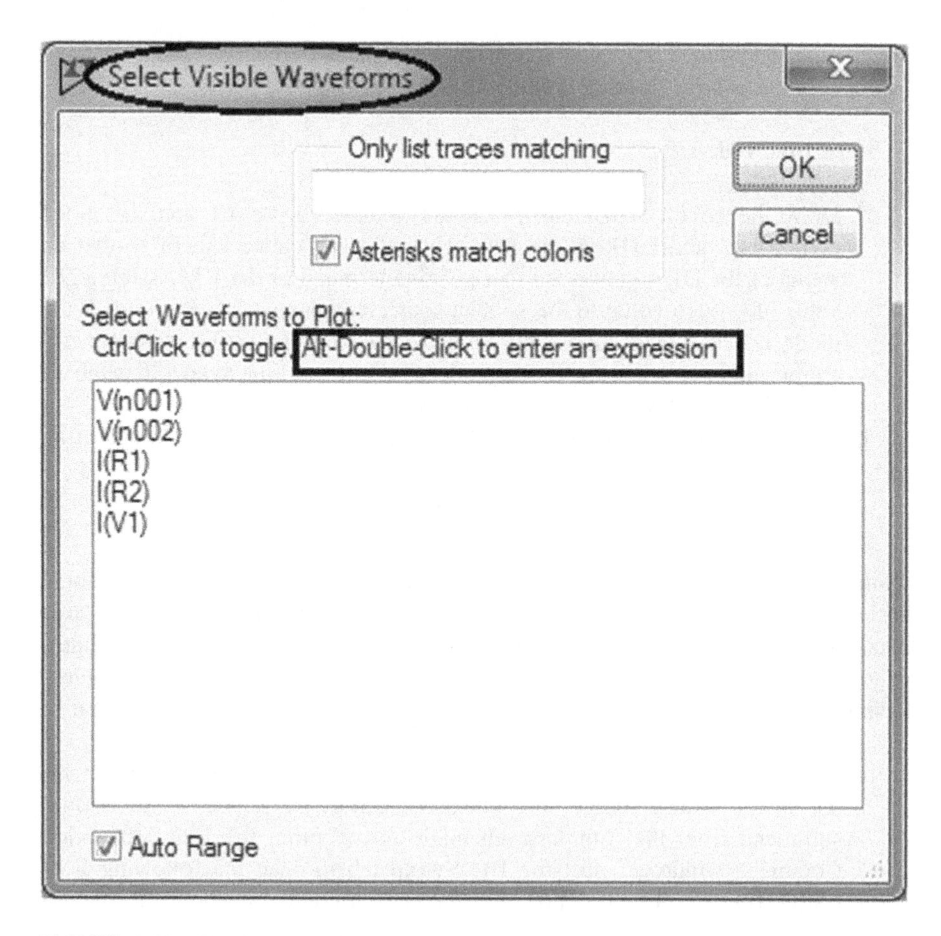

FIGURE 4.53 The Select Visible Waveforms Dialog Box

Alternatively, the **Visible Traces** option can be selected from the context-based menu (accessed by right-clicking somewhere on the waveform viewer window).

The **Select Visible Waveforms** dialog box is shown in Fig. 4.53.

Now, choose the desired circuit parameter (by left-clicking on the parameter/electrical quantity name and then OK) among the given parameters list from the given dialog box for plotting in the currently active waveform viewer window.

4.5 DC SWEEP TO PLOT I-V CURVE OF DIODE

A DC sweep is particularly useful for plotting dynamic I-V and transfer characteristic curves of semiconductor devices such as diodes, transistors, and many more for model verification by defining the parameters such as DC source-name, sweep type and sweep limits. A source-name is the name of the independent voltage source, current source, or circuit temperature to be swept.

Steps Involved:

i. Draw the circuit schematic. Place the voltage source V1 and the diode component model D1 on the new schematic and connect them so that an anode of the D1 is connected to a positive terminal of the V1. Assign a DC value of 0 (zero volts) to the voltage source component V1. Right-click the diode D1 body and choose the Pick New Diode option from its property editor window. Select the diode model 1N4148 from the available database in the LTspice XVII Component library.
ii. Label the Nodes using the Label Net schematic tool. Label the node at an anode terminal of the D1 as Vd.
iii. Save the drafted schematic circuit with a name diodedcsweep as shown in Fig. 4.54.

(**Note:** For diodes, BJTs, FETs, MOSFETs symbol, it is possible to specify some basic parameters or choose a specific manufacturer's model from the available database by clicking a **pick new** symbol_name **button** present on their attributes editor window (right-click to open it). Choosing the option brings up a window having the respective symbol's various models available in the Component library.)

iv. Here, perform a DC Sweep analysis and sweep a DC value of the independent voltage source component V1. Select the Edit Simulation Cmd sub-menu from the Simulate schematic menu. From the Edit Simulation Command window, select the **DC Sweep** tab to open the following **Edit Simulation Command** dialog box (Fig. 4.55) to set the sweep parameters as follows:

As the voltage ranges from 0 V to 1 V with an increment of 5 mV can be considered as most favorable for a silicon diode, type V1 in the **Name of 1st source to sweep** text box, select Linear from the drop-down menu next to the **Type of sweep** field, type 0 (minimum value) in the **Start value** box, type 1 (maximum value) in the **Stop**

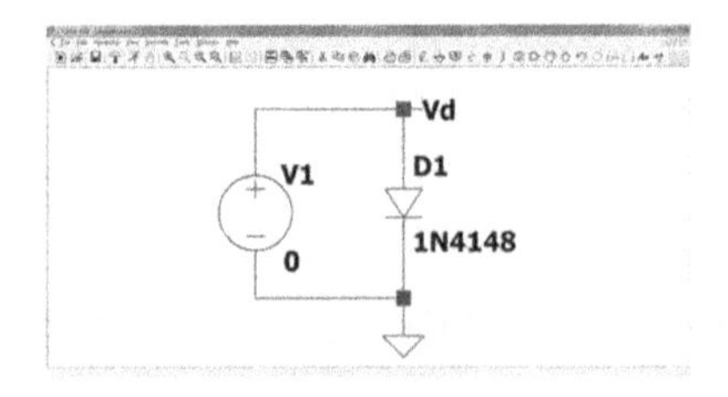

FIGURE 4.54 The Circuit Schematic.

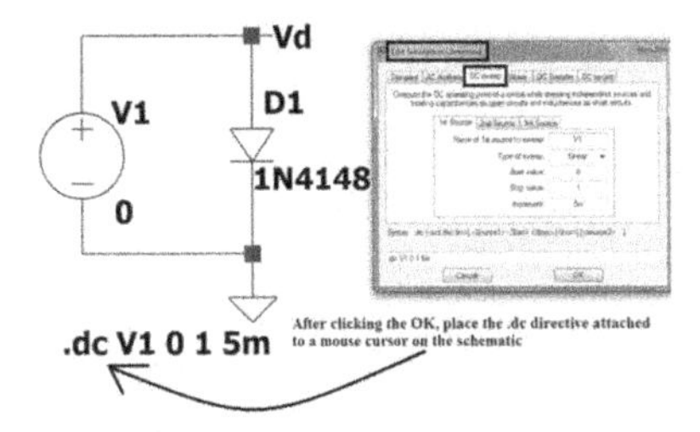

FIGURE 4.55 The Circuit Schematic with the .dc Analysis Statement.

value box and type 5m (step size) in the **Increment box.** After typing in all the inputs, click the OK and place the .dc V1 0 1 5m directive (automatically created by LTspice XVII based on the given inputs in the Edit Simulation Command dialog box) on the circuit schematic with a left-click as shown in Fig. 4.55.

(**Note:** To plot a reverse characteristic curve of a diode, set the DC voltage V1 to range from -150 V to 0 V with an increment of 5 mV.)

- **How to change the plotted graph's axis color:**

v. Before running a simulation, change the plotted graph's **axis** color to black by using the **WaveForm** tab under the Color Palette Editor dialog window as shown in Fig. 4.56.

To open the **Color Palette Editor** dialog, click on the Control Panel schematic tool, go to the Waveforms tab and select the **Color Scheme** button as shown in Fig. 4.56.

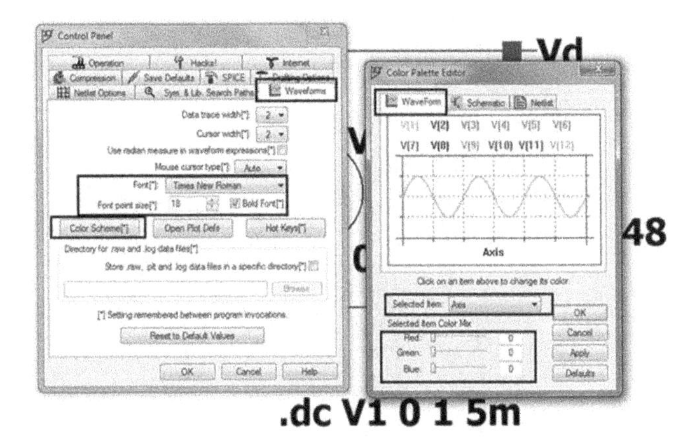

FIGURE 4.56 The Control Panel Waveforms Tab.

Also, change the waveform viewer axis font properties (select font type: Times New Roman, font size: 18, and bold) for better readability using the Control Panel schematic tool as shown in Fig. 4.56.

Click the OK button for executing the changes.

vi. Now, run the simulation. Move a cursor over the diode body and left-click to plot a current flowing through the D1. A change in current through a diode is plotted against variations in a diode voltage (Fig. 4.57).

Although the Vd can be varied at any arbitrary voltage, actually it is imperative to take into account that a real diode must adhere to its absolute maximum rating specified on the datasheet. Thus, the user may put any voltage across a diode while doing a simulation but practically in a real circuit build if power dissipation of the diode exceeds beyond the maximum limit, it can lead to smoke or the circuit may be damaged. LTspice XVII software does not give any warning and it is only the job of a circuit designer to adhere to the manufacturer's datasheet limits.

(**Note:** Correctness of a simulation depends upon the accuracy of the model.)

Remember: The user can also go for presenting the schematic and waveform viewer windows in a vertical manner by selecting the Window schematic menu and then clicking the Tile Vertically sub-menu)

- **Manipulating the Waveform Viewer:**
- **Using the .meas SPICE Directive Statement**

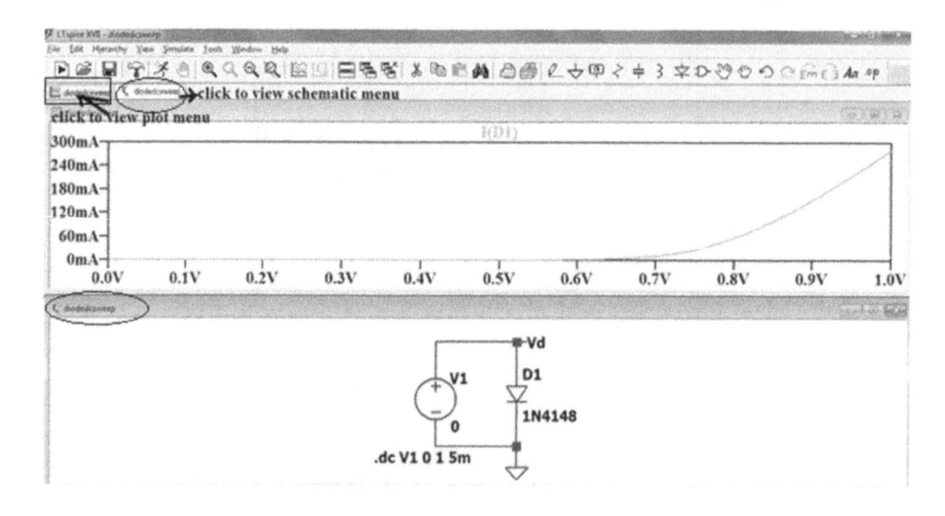

FIGURE 4.57 Simulating the Diode Current (an I-V curve).

vii. Measure quantities by placing the .meas directive on the schematic. To place the simulator directives, click on the schematic file icon to make it **active**. Now, select the SPICE directive schematic tool (the .op icon) or choose the SPICE Directive schematic option from the Edit pull-down menu. The action brings up the Edit Text on the Schematic dialog box as shown in Fig. 4.58.

First, ensure that the SPICE Directive radio button is checked or highlighted. Now type a needed command, that is, the desired directive (empty or complete statement), and then left-click to place the directive attached to a cursor onto the circuit diagram.

Remember: If the statement is too long, press Ctrl + M to start a new line.

Thus, type the desired directive, say .meas in the Edit Text on the Schematic box and place it on the schematic after pressing the OK button. The directive statement can further be accessed easily for editing purposes by right-clicking on the placed directive as shown in Fig. 4.58.

Thus, to know a current (through the specific device model or node) or voltage (at the particular node) when a specified condition is reached, type the desired directive, say .meas in the Edit Text on the Schematic box and click the OK. Place the .meas directive (that is now attached to a cursor) on the schematic. Now, access the already placed directive for editing purpose by right-clicking on it. The action brings up the **.meas Statement Editor** dialog window as shown in Fig. 4.58.

Choose the entries, say **Genre:** FIND and **Measured Quantity:** I(D1), to find the Id (a required result name) at a point when V1 = 0.8 V. After choosing the FIND option from the pull-down next to the Genre field, a field for entering a quantity (ordinate data or dependent variable, say I(D1) here) to be measured appears

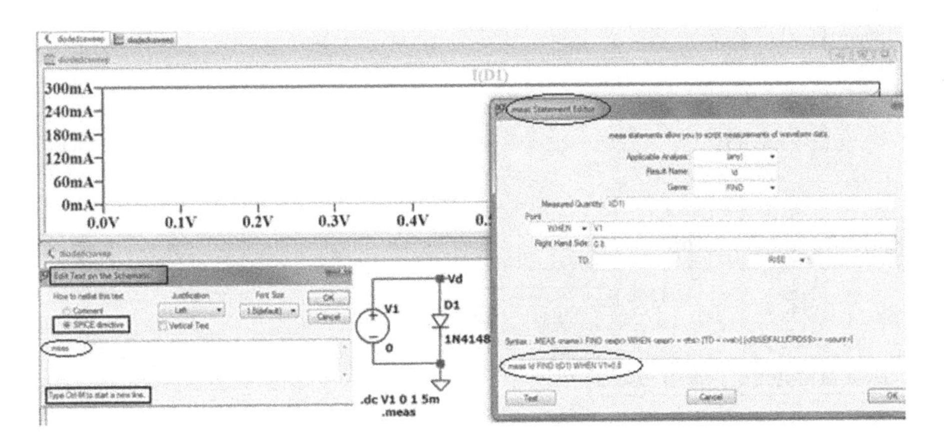

FIGURE 4.58 Editing the .meas for Completing the Statement.

followed by two more fields to specify the condition(s), say when V1 = 0.8 V, for which an output data value, say Id, needs to be determined.

If the syntax of the .meas directive statement (used to print a data point value/expression at a specific point or when a condition is met or a requested category over a range of abscissa) carries a value in angle brackets < >, it is mandatory and must be specified. The pipe | symbol tells to choose one or the other value and the plus + symbol represents a continuation of the statement on the next line. The first field of .meas statement includes many options such as [(any)|AC|DC|OP|TRAN|TF|NOISE] and is named Applicable Analysis. For circuits involving AC or transient or noise calculations in the time or frequency domain, the first field is relevant and should be selected. Then, follows a mandatory result name in angle brackets, followed by a category for evaluation (one point measurement or operation over an interval) which includes many options. If no category is specified, the result represents a distance along an abscissa (independent variable plotted along a horizontal axis) between the TRIG and TARG points.

The **Test** button [Test] on the bottom left-hand side evaluates the result.

After completing all the entries for the fields provided in the **.meas Statement Editor** dialog box, click the OK. Ensure the complete statement is placed on the schematic. Now, right-click on the placed directive statement and click the **Test** button to find the diode current when the diode voltage is 0.8 volt as shown in Fig. 4.59.

The diode current comes out to be 0.0571757 A or 57.1757 mA when a voltage across the diode is taken to be 0.8 V.

Thus, it can be said that the .meas SPICE directive statements are used for evaluating user-defined electrical quantities. Among the two basic types of .meas directive statements available, the first one **refers to a single point along the abscissa** (an independent variable plotted along a horizontal axis, for example, a

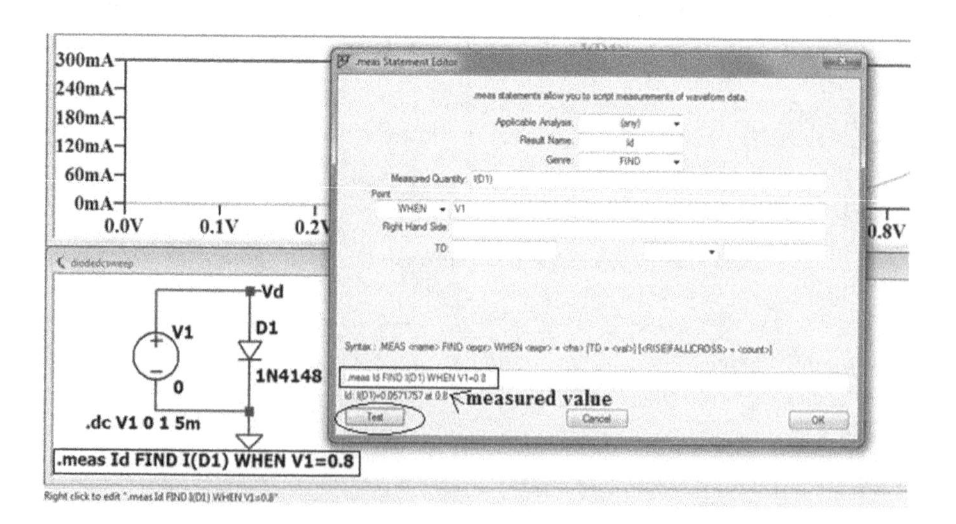

FIGURE 4.59 The Results Displayed in the .meas Statement Editor Window.

time axis of the .tran analysis) and the other **refers to a range over an abscissa.** The first version is used to **print the data value or expression at a specific point on an abscissa or when a specific condition is met, for example,** finding a value of the dependent variable y when the x occurs. It is to be noted that while referring to one point along the abscissa, the requested result is based on an ordinate data y and if no ordinate information is requested by the user, then the .meas statement prints point on the abscissa that the measurement condition occurs.

The other version referring to a **range over the abscissa** includes a rise and fall time, time delay, average, RMS, minimum and maximum value, peak-to-peak, integrals over an interval. The range over the abscissa is specified with the points defined by **TRIG** and **TARG**. If omitted, the TRIG point defaults to the start of a simulation and the TARG point defaults to the end of a simulation data. If the TRIG, TARG, and the previous WHEN points all are omitted, then the .meas statement operates over an entire range of the data. If no measurement operation is specified, a result of the .meas directive statement is a distance along an abscissa between the TRIG and TARG points.

- **Using the LTspice Numbered Cursors**

vii. Read the data point values using the LTspice numbered cursors. LTspice XVII allows the waveform viewer's attached numbered cursor (the Cursor 1 or Cursor 2) feature to read horizontal axis and vertical axis values of a particular point on the plotted graph. By attaching two cursors to a trace, the user can measure a displacement, time delay (or phase difference), slope, maximum and minimum values, impedance measurement, rise and fall times, etc. The cursor(s) can be dragged along the plot and they follow the waveform. When the cursor is attached, the floating cursor window (or data window) pops up providing amplitude/magnitude levels, time, and frequency information.

Right-clicking on the waveform node-title label/name at the top of the waveform viewer graph window brings the Attached Cursor drop down box where the user can select to attach a single cursor (the 1st or the 2nd) or both the cursors (the 1st & 2nd) to set a cursor cross-bar along with popping-up a new floating readout display window with the cursor position information. By moving a mouse pointer over the new cross-bar and left-clicking when a number 1 appears, the attached cursor can be dragged with a mouse (or moved with the cursor keys) along the graph to change the position/location of the numbered cursor so that the desired cursor position information can be read in the floating cursor window.

- **Attaching the numbered cursor:**

First, delete the .meas directive statement using the **Cut** tool and uncheck (or un-tick) the **Grid** check box. To attach the Cursor 1, first right-click over the name of the traced waveform, that is, **I(D1)** to open the **Expression Editor** window, and then left-click on the **1st** scroll-down menu next to the **Attached Cursor** field (see

Fig. 4.52). The action brings up the cross-hairs on the plot pane along with the floating cursor window (little data window) displaying the initial location of the Cursor 1, that is, the position where it is placed initially by default (see Fig. 4.60).

Now, move a mouse over the cross-hairs (where the Cursor 1 is placed by default) to see the yellow-colored Cursor 1 as shown in Fig. 4.60:

Hold a left-mouse button down and drag the yellow-colored Cursor 1 left or right to observe a horizontal (x-axis) and a vertical (y-axis) value of a specified point. Here, the diode current is observed at the diode voltage of 0.8 V by dragging the Cursor 1 to the right till a text box next to the Horz field in the little cursor window shows 0.8 V or 800 mV. Note down a value in a text box next to the Vert field in the cursor window for the operating diode current value (y-axis value). A current through the D1 can be read as 56.798507 mA at the diode voltage of 799.4769 mV or 0.8 V (see Fig. 4.61).

Also, the cursor position is labeled by selecting the **Plot Settings -> Notes and Annotations** -> Label Curs. Pos. under the plot menu diodedcsweep as shown in Fig. 4.61.

- **Using the above Expression Editor window to plot expressions on traces**

viii. In the **Enter an algebraic expression to plot** field box (see Fig. 4.52), the user can type the expression V(vd)*I(D1) to measure a dissipated power or loss (a diode current multiplied by a diode voltage drop) across the diode D1 and click the OK to see the plotted graphs.

4.5.1 Operating Point Establishment

- Steps for locating a DC Operating Point:

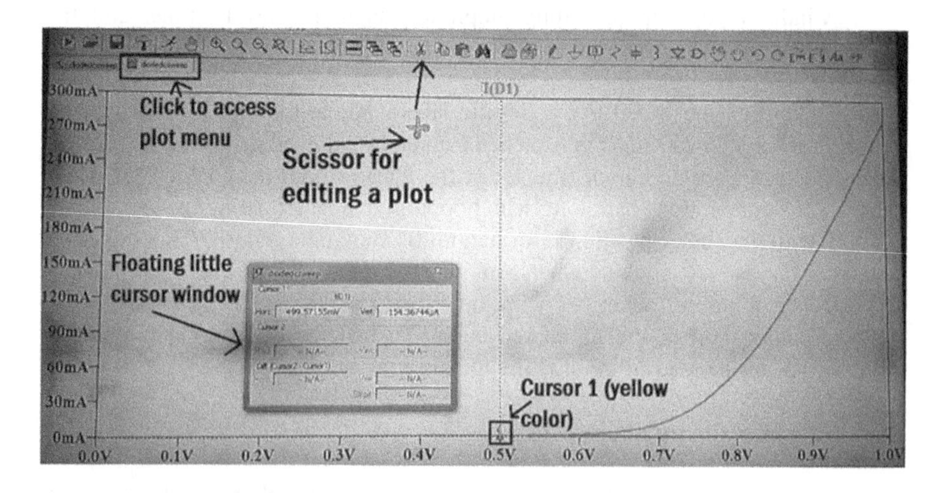

FIGURE 4.60 The Attached Cursor 1 and its Data Window.

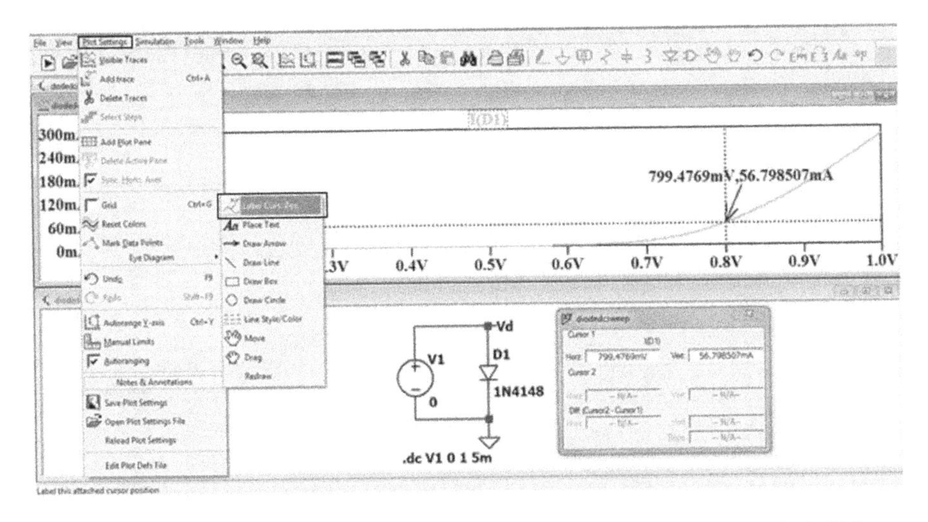

FIGURE 4.61 The Cursor Read-Out Display for the Diode Current at 0.8 V Diode Voltage.

ix. The target is to choose an operating point for the device so that when it is used in the circuit, the diode should operate at this point. Let, a supply voltage is taken to be 1 V and a voltage across the diode is selected to be **0.8** V (fixed operating point voltage) when connected to the circuit. First, click on an empty area in the plot pane (or click the waveform file icon diodedcsweep) to view the **Plot Settings trace** menu . Go to the Notes and Annotations -> Draw Line under the **Plot Settings** waveform viewer menu as shown in Fig. 4.62:

Draw a vertical line from the point on an x-axis (x-intercept) at 0.8 V (the restricted operating point diode voltage). The next step is to draw a line from an x-intercept at the supply voltage point, that is, 1 V, through the operating point (or Q-point), all the way to a y-axis. This is the DC load line along with which the circuit always operates.

- **Finding the Resistance Value for Establishing the Diode Voltage and Current:**

x. The first step is to compute a total amount of resistance necessary for the circuit to follow this line by dividing an x-intercept (maximum voltage value) by a y-intercept (maximum current value). A y-intercept or the maximum circuit current can be computed using the Cursor 1 (left-click on the trace label I(D1) to automatically attach a single cursor) as shown in Fig. 4.63.

Place the Cursor 1 at the position where a text box next to the Horz field in the little cursor window shows 1 V (the selected value of an x-intercept) to read that a y-intercept is here 281.98 mA.

(**Note:** The already traced plot can be deleted by right-clicking on the trace/plot label to see the results of a new simulation.)

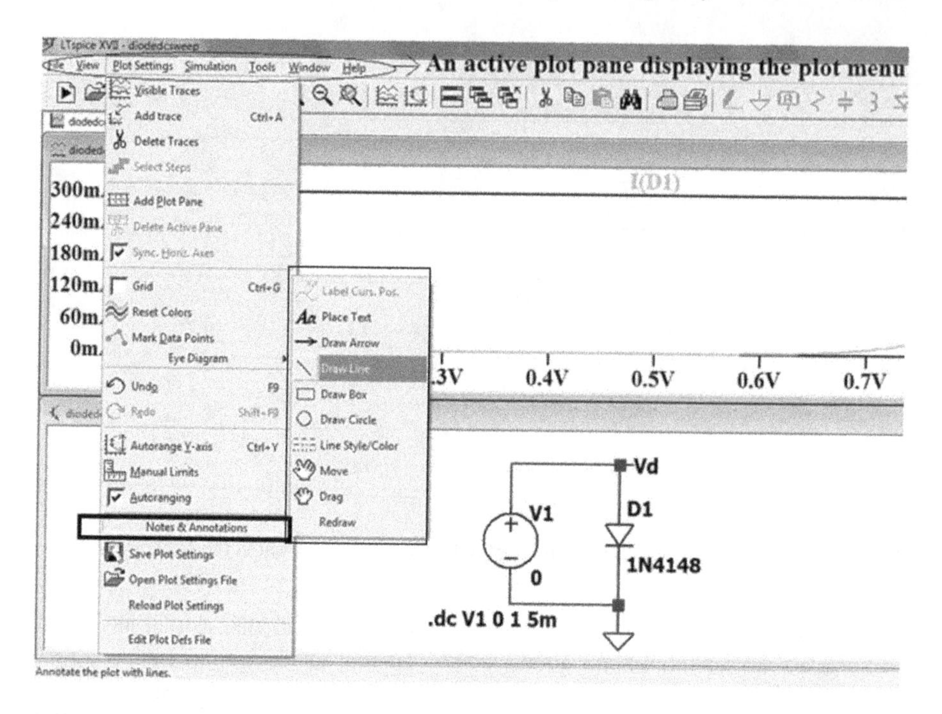

FIGURE 4.62 The Plot Settings Plot Menu.

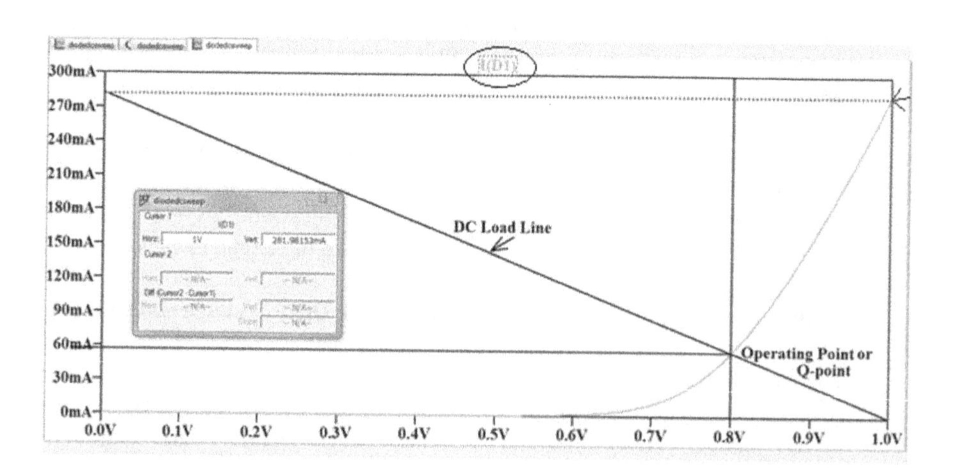

FIGURE 4.63 Locating the Diode Current at a Fixed Diode Operating Voltage.

A y-intercept can also be computed using the .meas directive statement in LTspice XVII with the entries, say Genre: FIND and Measured Quantity: I(D1), at a point when V1 = 1 V and it displays the answer to be equal to 281.98 mA. The **Test** button [Test] evaluates the result.

The resistance required in the circuit to follow the DC load-line is given by the ratio of x-intercept (voltage) to y-intercept (current), that is, 1V/281.98 mA, which gives **3.55** ohms (Ω).

Similarly, an operating point diode current can be read by moving the Cursor 1 to a location where a text box next to the Horz field in the little cursor window shows 0.8 V or 800 mV (the selected operating voltage for the diode). Note down a value in a text box next to the Vert field in the cursor window for the operating diode current value (y-axis value). A current through the diode when conducting at the operating point voltage equal to 0.8 V can also be computed using the .meas statement and it is equal to 57.1757 mA.

xi. Results of the waveform viewer window can be saved by selecting the Save Plot Settings option from the Plot Settings trace menu as shown in Fig. 4.64.

Click the Save button.

The saved plot file can also be opened by choosing the **Waveforms** option from the scroll-down menu on the extreme right of the **Open an existing file** dialog window as shown in Fig. 4.65.

Now, select the resistor component and assign the computed value, that is, 3.55 Ω and draft the circuit using the diode as a circuit element as follows:

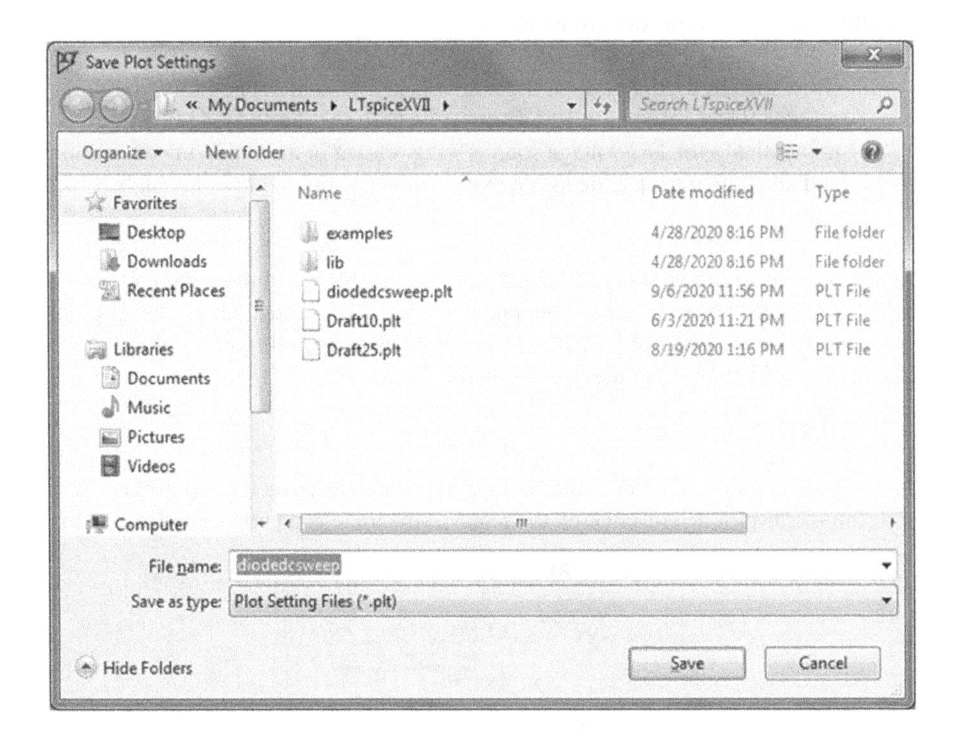

FIGURE 4.64 The Save Plot Settings Window.

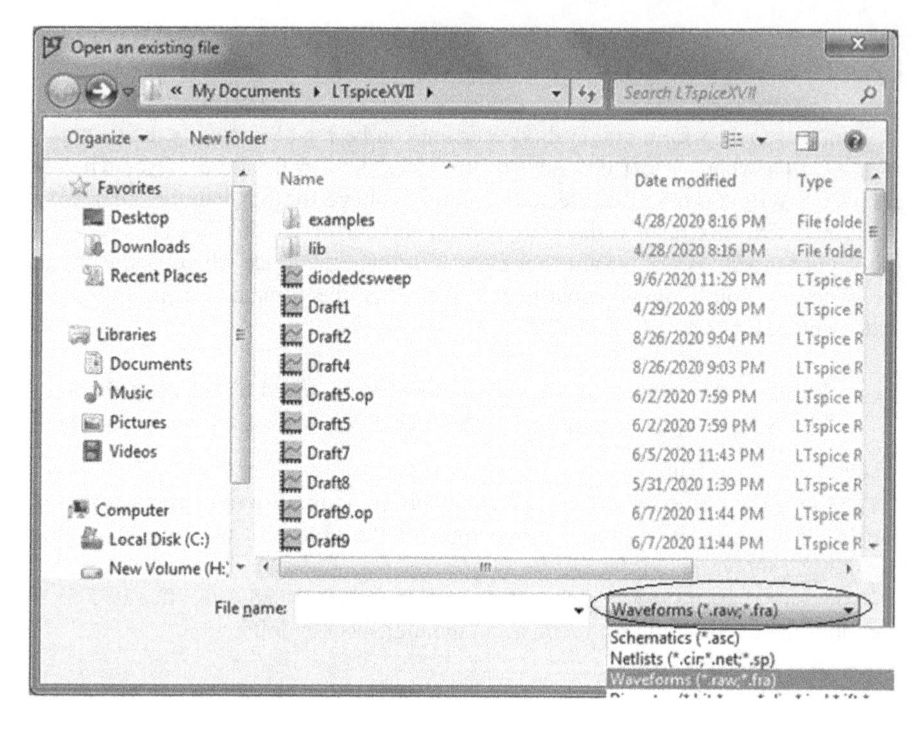

FIGURE 4.65 The Open An Existing File Window.

xii. Draw the circuit diagram on the new schematic having a combination of the resistor R1 of resistance 3.55 Ω in series with the diode 1N4148 connected in parallel to the DC voltage source V1 as shown in the following discussion and label the output node as Vd. Now, run a DC operating point analysis for verification.

As can be seen from Fig. 4.66, the diode voltage is 0.799 V and the circuit current is 56.57 mA. This value is almost identical to the actual location of the Q-point specified by the user on the load line in Fig. 4.63. The operating point of the diode (previously selected by the user) is verified.

Few Tips:

- The Expression Editor window is used to delete an active (present at the moment) plot. Right-click on the name of the trace and choose the **Delete this**

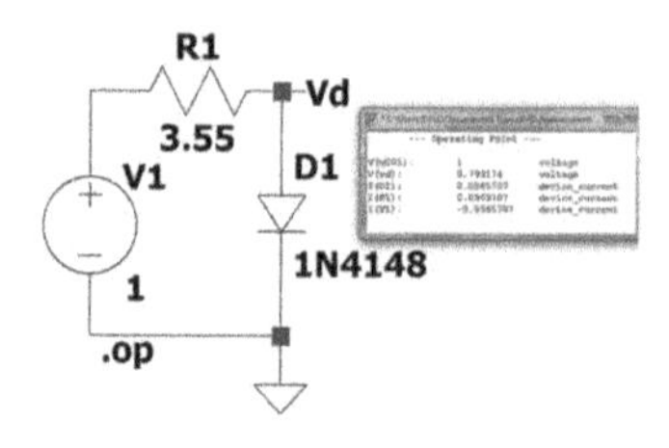

FIGURE 4.66 The .op Analysis Simulation Results.

Trace option from the Expression Editor window. Click the OK to delete a current trace.

- Any directive statement once placed can be edited by right-clicking on the directive statement on the schematic.
- If something went wrong while simulating the drafted schematic, a spice error log is created which can be viewed by selecting the SPICE Error Log sub-menu under the View menu or by using shortcut keys Ctrl + L.

4.6 DC SWEEP FOR OUTPUT CHARACTERISTIC CURVE OF BJT

i. Draw the schematic as shown in Fig. 4.67. Search for an NPN bipolar transistor in the component library and place (rotate before placing if required) the 2N3391A NPN transistor model for simulation as shown in the diagram. Connect its base terminal to the current source I1 and a collector terminal to the voltage source V1. Set values of the I1 and V1 to be equal to 0 in their respective value window.

(**Note:** A resistor can also be added on the left of the circuit for reducing an input base current to microampere levels in Fig. 4.67.)

- **Searching the 2N3391A NPN Transistor Model:**

ii. Right-click on the NPN transistor Q1 (with a symbol name npn) to open the component value editor window (Fig. 4.68) so that we can access the database of models associated with it. Now, click on the **Pick New Transistor** to configure the NPN transistor model to manufacturer's attributes by selecting any of the available models in the **Select Bipolar Transistor** dialog window.

By default, the simulator is installed with a list of transistor models as shown in Fig. 4.69 from which any of the component model names (on the extreme left of the window) can be selected for use in the circuit drawing. Left-click on the 2N3391A NPN transistor model (Fig. 4.69) to select it for simulation and click the OK.

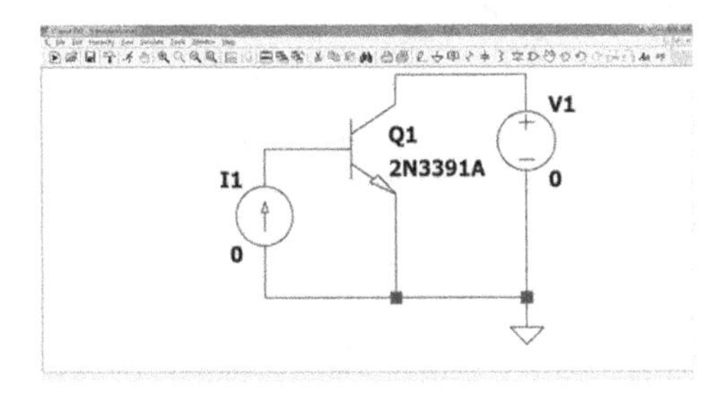

FIGURE 4.67 The Transistor Circuit Schematic.

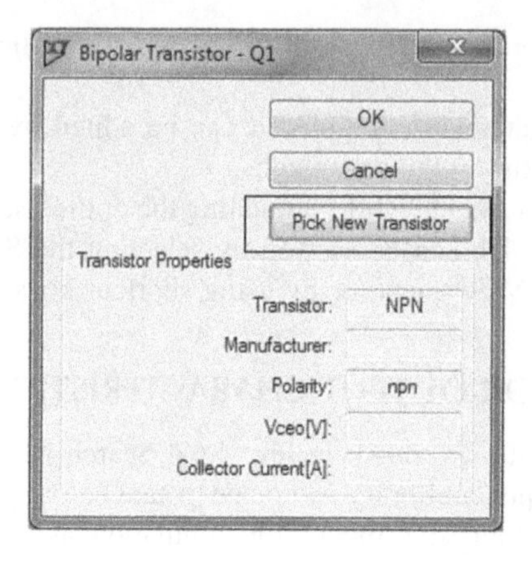

FIGURE 4.68 The Bipolar Transistor Dialog Window

Select Bipolar Transistor

OK

Cancel

Part No.	Manufacturer	Polarity	Vceo[V]	Ic[mA]	SPICE Model
2N3904	NXP	npn	40.0	200	.model 2N3904 NPN(IS=1E-1
FZT849	Zetex	npn	30.0	7000	.model FZT849 NPN(IS=5.85
ZTX1048A	Zetex	npn	17.5	5000	.model ZTX1048A NPN(IS=1
2N4124	Fairchild	npn	25.0	200	.model 2N4124 NPN(Is=6.734
2N3391A	Fairchild	npn	25.0	500	.model 2N3391A NPN(Is=12
2N5089	Fairchild	npn	25.0	100	.model 2N5089 NPN(Is=5.911
2N5210	Fairchild	npn	50.0	100	.model 2N5210 NPN(Is=5.911

FIGURE 4.69 Selecting the Component Model from the Library.

- **Choosing an Analysis Type:**

Here, a DC Sweep analysis is used to sweep both the independent sources, that is, the output voltage source V1 and input current source I1. The DC voltage is typically swept in a smooth linear fashion while the current source is stepped with significant microampere stages.

iii. Select the **Edit Simulation Cmd** sub-menu from the **Simulate** schematic menu to open the following **Edit Simulation Command** dialog box (Fig. 4.70). From the Edit Simulation Command window, select the **DC Sweep** tab to set the sweep parameters as follows:

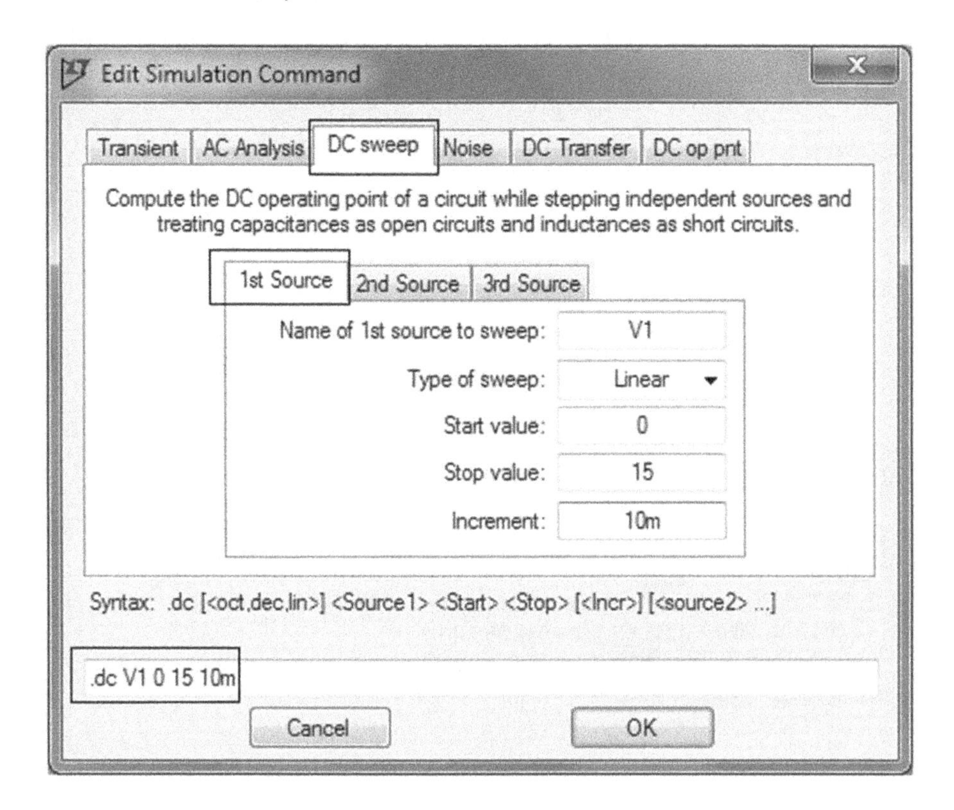

FIGURE 4.70 The Edit Simulation Command Dialog Box.

First, perform the DC sweep on the output voltage source V1 (1st source between collector and emitter) and set the parameters to range from a minimum of 0 V to a maximum of 15 V with an increment of 10 mV as shown in Fig. 4.70.

The earlier-mentioned action creates a smooth linear output voltage (collector-emitter voltage) rise.

Now, click the **2nd Source** tab in the Edit Simulation Command dialog window to edit the values for the second input (base) current source I1. Increment the DC current source I1 (2nd source) from a minimum value of 0 A to a maximum value of 100 μA with a step size equal to 10 μA as shown in Fig. 4.71.

After completing the given two settings, click the OK to place the complete analysis statement on the schematic.

(**Note:** Smaller steps take longer for it to simulate but produce more accurate output. Larger steps take less time, but do not provide a good representation of data.)

- **Running the Simulation**

iv. After placing the .dc SPICE directive statement for a DC sweep analysis, press the Run schematic tool button to start the simulation. Move a cursor to

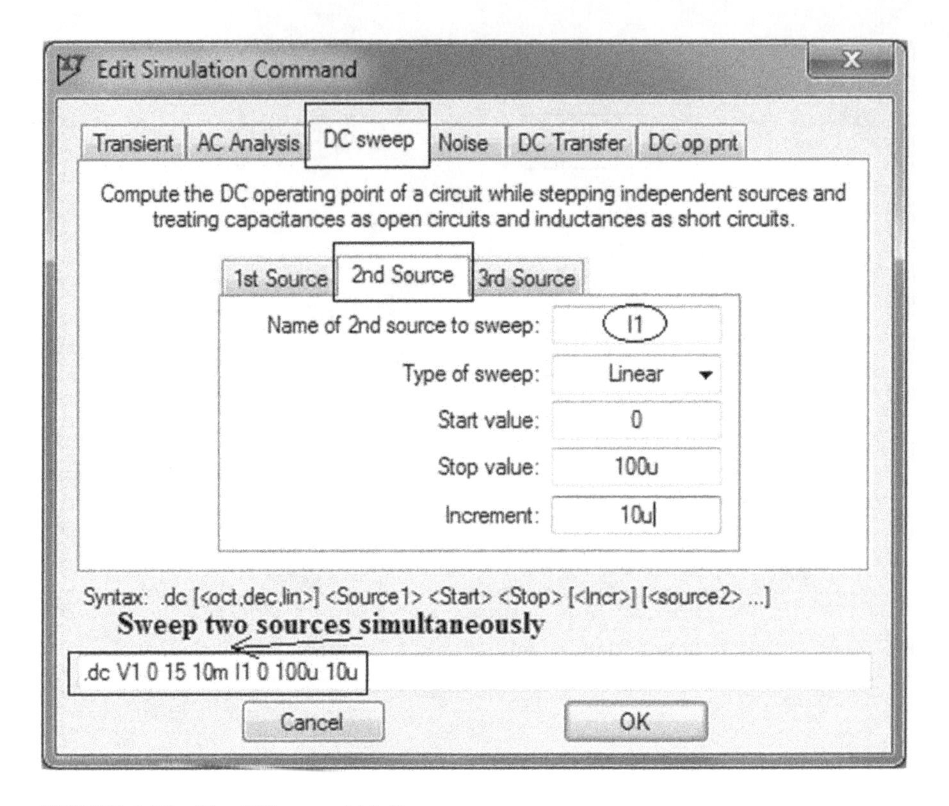

FIGURE 4.71 The DC sweep Tab Parameters.

the **collector pin** of the transistor Q1 and place it on the collector node (**endpoint**) and left-click when the **current sensor** appears as shown in Fig. 4.72 to plot the collector current.

The following output I-V characteristic curves are obtained as shown in Fig. 4.73:

The command .dc I1 0 100 u 10 u varies the input base current from 0 to 100 μ with a step size of 10 μ, creating 11 simulations with separate values (0 μ, 10 μ, 20 μ, 30 μ, 40 μ, 50 μ, 60 μ, 70 μ, 80 μ, 90 μ, and 100 μ). The plot pane shows the collector current for the 11 simulations run (in different colors by default, where the first trace color is green).

(**Note:** For plotting a current into a pin of the component with more than two terminals, first place a cursor over the pin of interest so that the current probe sensor appears, and afterward left-click the pin of interest.)

The transistor output current is measured in a range of mA (milliamperes). The transistor characteristics typically change (although follow the same pattern) based on collector, emitter, or base measurements (voltage and/or current) and the current gain beta.

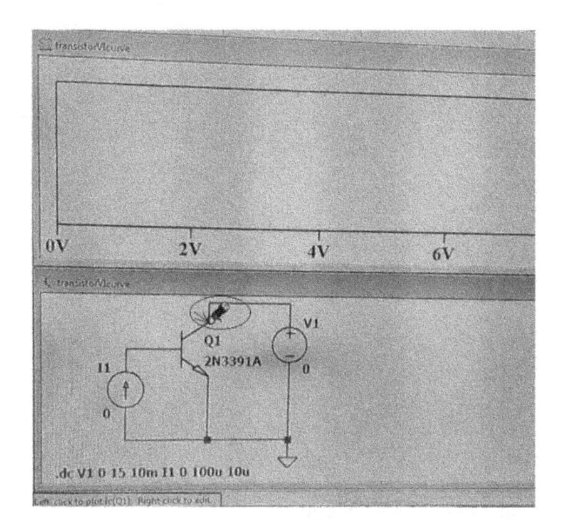

FIGURE 4.72 The Current Sensor Probe

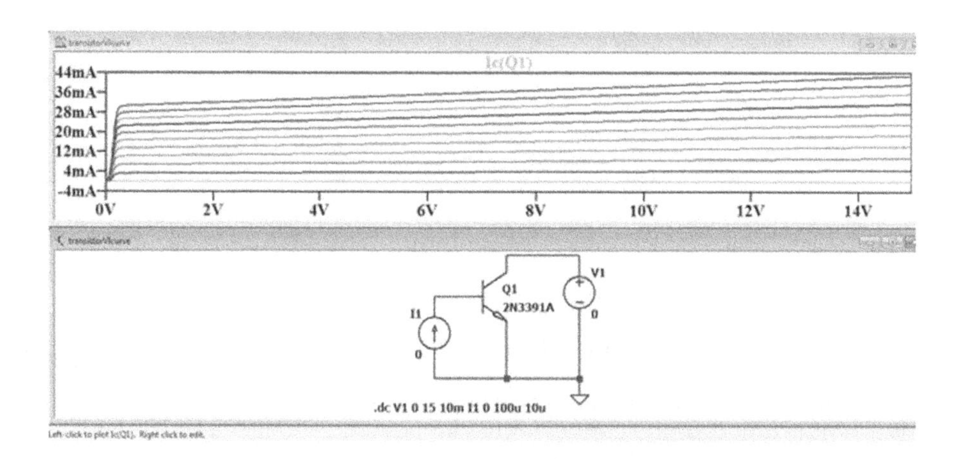

FIGURE 4.73 The Output I-V Characteristic Curves

- **Inserting Legends:**

v. To add legends to the trace for each simulation run signifying the assigned base current values, right-click on the given waveform viewer window or trace/plot pane to open the context-based menu, and afterward go to the **View -> Step Legend** as shown in Fig. 4.74.

The Legend dialog window appears on the screen which should be expanded by pulling it from the rightmost corner to obtain the expanded legend box to see all the legends as shown in Fig. 4.75.

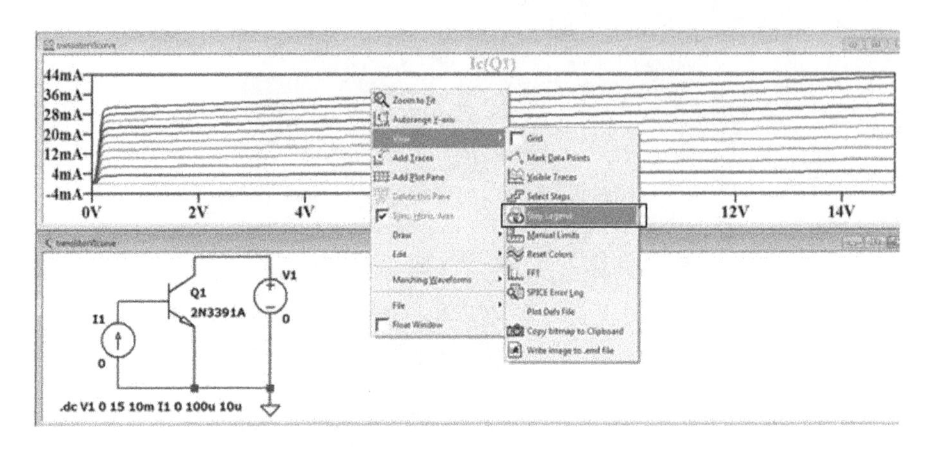

FIGURE 4.74 The Step Legend Sub-Menu

Legend

I1=0 (Run: 1/11)
I1=10μ (Run: 2/11)
I1=20μ (Run: 3/11)
I1=30μ (Run: 4/11)
I1=40μ (Run: 5/11)
I1=50μ (Run: 6/11)
I1=60μ (Run: 7/11)
I1=70μ (Run: 8/11)
I1=80μ (Run: 9/11)
I1=90μ (Run: 10/11)
I1=100μ (Run: 11/11)

FIGURE 4.75 The Legend Window (Expanded View)

4.6.1 How to Locate Transistor Operating Point

For a given value of collector-emitter voltage, the value of collector current can be computed as follows:

vi. Let, the device is set up to operate at a bias point voltage equal to 7 V (the collector-emitter voltage of the transistor) so that regardless of the input signal swing, the transistor remains in an active region and produces faithful amplification. Draw a vertical line starting from a point at which the collector-emitter voltage is 7 V. The next step is to draw a DC load line (a limitation laid on voltage and current in a nonlinear device by an external circuit) starting from an x-intercept or cutoff point at the supply voltage point (say 15 V as is taken here, the maximum possible collector-emitter voltage), passing through the Q point or operating point (say 7 V collector-emitter voltage as is selected here for its linear operation) and extending up to y-axis, along which the circuit always operates (Fig. 4.76).

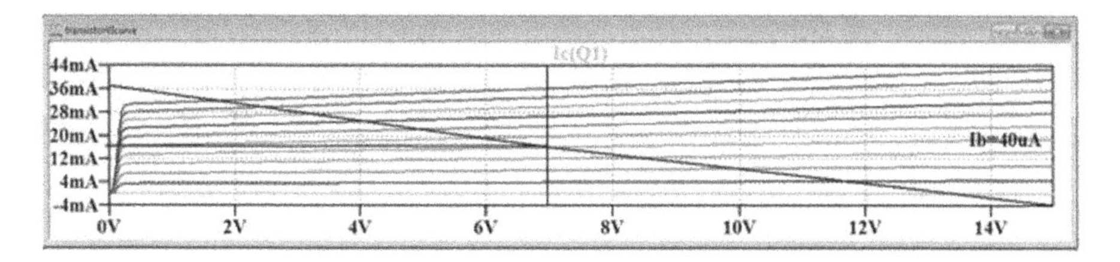

FIGURE 4.76 Locating the Operating Point

A y-intercept or saturation point (i.e., the maximum possible collector current where the collector-emitter voltage is zero) for the case when the base current value is fixed at 40 μA (microamperes) and the operating point collector-emitter voltage is limited to 7 V, is approximately 36.5 mA.

Now, for the case when a base current value is fixed at 40 μA (i.e., the fifth step output curve), it can be observed by drawing a horizontal line (from a point of intersection of the load-line and the curve) extending up to y-axis that the transistor circuit collector current floats around 15 mA. To find an exact value, use the .meas directive statement as shown in Fig. 4.77.

Thus, at the fifth step output curve (see Fig. 4.77) when the operating point collector-emitter voltage is limited to 7 V, the operating collector current is limited to 15.9112 mA. Now, the total amount of collector resistance necessary for the

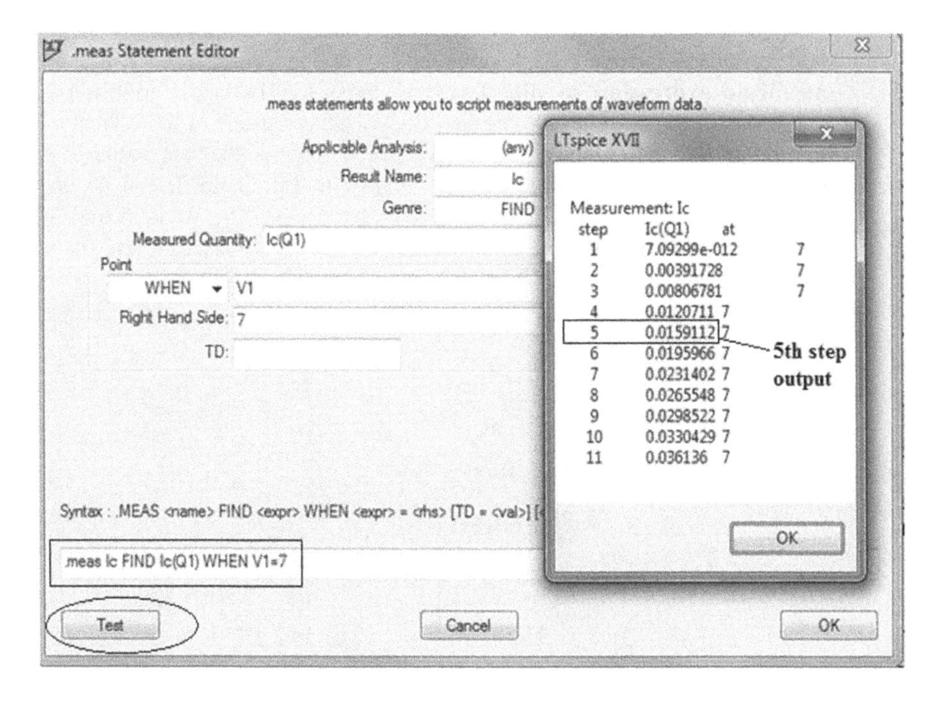

FIGURE 4.77 The .meas Statement Editor Window

circuit to follow the restricted load line can be calculated by dividing an x-intercept (the supply voltage connected across the collector and emitter), say 15 V by a y-intercept (maximum collector current), say 36.5 mA. Therefore, the required collector resistance is given by 15V/36.5 mA = 410.86 Ω.

- **Verification of the restricted operating point of the circuit with the computed collector resistance value:**

vii. Now draw the circuit as follows and run an operating point analysis when the collector resistance value is chosen to be 410.86 Ω. The following output appears (Fig. 4.78):

The collector-emitter voltage can be read as 8.27 V and the collector current is 16.375 mA. Thus, the .op results are quite close to the operating point selected.

4.6.2 Doing Math for Computing Beta (and Alpha)

- **Plotting the Transistor DC Beta (CE current gain):**

viii. In the circuit schematic shown in Fig. 4.67, change the current source value to 40 microamperes. Now, edit the .dc command to sweep only the voltage source V1 (i.e., place the directive .dc V1 0 15 10m). To plot beta (the current gain of the transistor) against the V1, left-click on the Pick Visible Traces schematic icon (Pick Visible Traces becomes active after running a simulation) and press Alt + Double-click to open the Expression Editor. In the **Enter an algebraic expression to plot** text box, type Ic(Q1)/Ib(Q1) to observe variations in the current gain (beta) values with variations in the collector-to-emitter voltage. Also, attach the Cursor 1 to find the beta value at a specified voltage in the little Cursor window that pops-up. It can be observed that the transistor DC beta value is 388.5 when the V1 is at 6 V. It can be observed that the transistor beta floats around **425** after the transistor output current stabilizes.

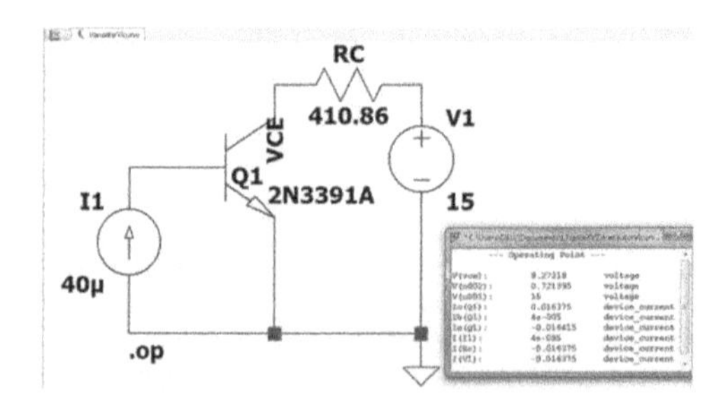

FIGURE 4.78 The .op Analysis Simulation Results.

(**Note:** It is possible to enter a mathematical expression through the Visible Traces and Add Traces options using all the signals present in the circuit. Also, by right-clicking on the label of a trace already plotted, the Expression Editor dialog window appears for entering a mathematical expression.)

- **Plotting the Transistor DC Alpha (CB current gain):**

ix. To plot alpha, type Ic(Q1)/-Ie(Q1) in the **Enter an algebraic expression to plot** text box. Here, a minus sign is used to get positive values as the direction of Ie is opposite to that of the Ic.

4.6.3 Temperature Effects on the Transistor I-V Curve

LTspice XVII provides an already defined global parameter for temperature named temp. Before observing how temperature affects the drafted circuit, ensure that some temperature-dependent components must be there in the circuit.

To sweep temperature and observe the effects, type and place the following directive statement:

.step temp 25 100 50

The given .step directive command raises a variable temp from 25 to 100 degrees Celsius in fifty-degree increments. The default unit is degree Celsius.

If only the .step is placed on the schematic. Now to complete the statement, right-click on .step to open the .step Statement Editor window and type <global temp> in the Name of parameter to sweep field. Select the nature of sweep, start value, stop value, increment value, and click the OK to place the complete .step directive statement. Run a DC sweep simulation by sweeping only the transistor Q1 collector voltage and keep the base current at 40 μA to see the effects of temperature as shown in Fig. 4.79.

4.6.3.1 Thermistor Modeling in the Circuit

xi. For observing the effects of the component parameter value that changes with temperature, for example, to model a thermistor:

- Add the resistor R1 in the given transistor circuit schematic as shown in Fig. 4.80.
- Replace the R1 component value with a parameter name encased in parentheses, for example, replace the resistance parameter value of the resistor model with {therm}.
- To put in the resistor model an equation that describes the resistance as a function of temperature, place the directive statement .param therm=(temp*0.5)+10 on the schematic (here, thermistor follows a linear relationship with temperature). Thus, the .param statement can be used for storing values in a user-defined variable.
- Now, enter the .step simulation command for sweeping the parameter temp (already defined in the simulator) using the SPICE Directive schematic tool,

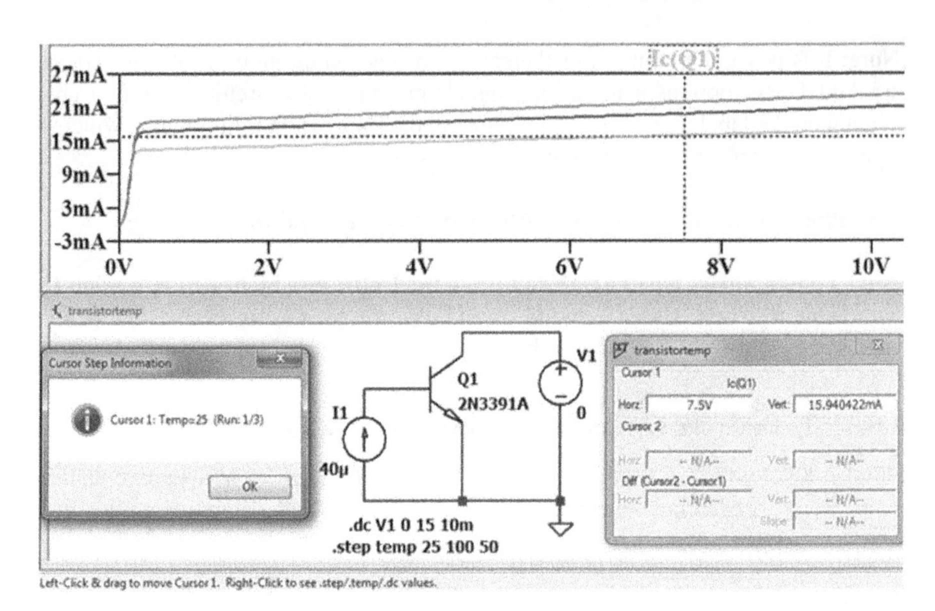

FIGURE 4.79 Temperature Effects on I-V Curve.

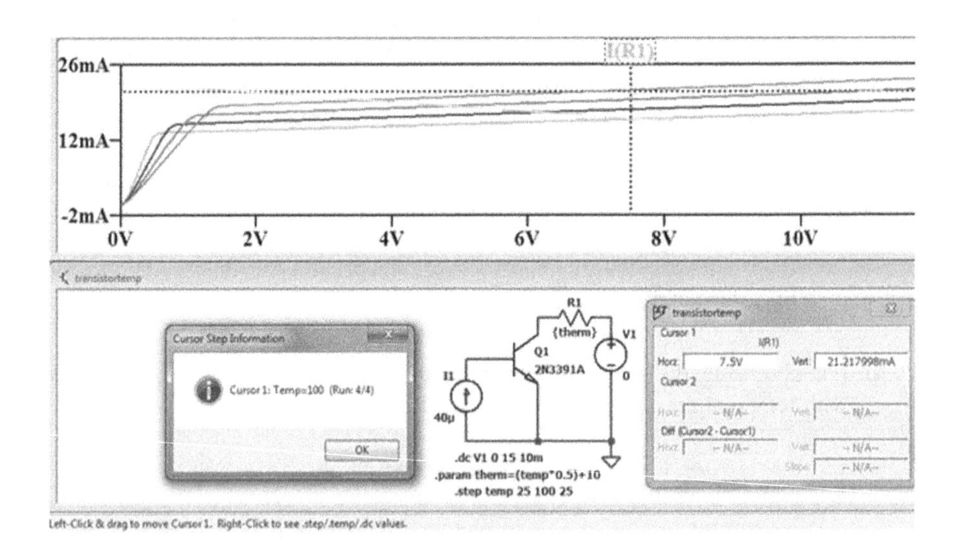

FIGURE 4.80 Modeling a Thermistor.

for example, .step temp 25 100 25. A directive statement is a text which gets included in the netlist. The .step command editor where the parameter to be swept followed by start, stop and incremental values can be entered is accessed by right-clicking on the blank .step statement.

- The parameter value may depend upon any other parameter such as the circuit voltage or current.

Here, rotate the R1 only once to orient the model in agreement with the desired current direction.

(**Note:** LTspice XVII does not provide an indication of which component terminal is the positive one. It makes the direction of a current difficult to be determined, but the accuracy of LTspice XVII is not at all affected and only a phase response of the plotted current curve may change by 180 degrees depending on the orientation of the component.)

4.7 SIMULATING VOLTAGE-CONTROLLED VOLTAGE SOURCE (VCVS)

- Steps Involved:

i. To construct the circuit schematic having the VCVS connected to the resistor, the user needs to place the component VCVS on the schematic.

- Searching VCVS
  The user should refer to the Help schematic menu for searching detailed information on the VCVS component (such as the component symbol name, etc.) by selecting the Help Topics sub-menu option. Help Help Topics F1 In the Search tab, after typing voltage source click on the List Topics button and select the **E. Voltage-Dependent Voltage Source** from the **Select Topic to Display** scroll-down menu.
  Alternatively, go to the Contents tab Contents Index Search, then double-click on LTspice XVII LTspice XVII. Now, double-click on Introduction Introduction, then on LTspice®, and afterward on Circuit Elements. Now, double-click on the respective component to know the symbol names.
  Remember: If we end up in a sub-library, just scroll to the left in the component selection window and double-click on a symbol [..], that is, the icon [..] present on the top to **come back** to the main library.
- **Placing the VCVS across the Resistor**

First, select the Component icon Component from the schematic toolbar, and then choose the e2 component symbol (or type a letter e2 in the text box of the Select Component Symbol dialog window). The component symbols are organized in directories.

It is to be noted that both letters e and e2 are the designations for the VCVS component which differ only in terms of their layout or orientation. The VCVS has two control input lines denoted by the symbols + and – (see Fig. 4.81).

For the dependent source component VCVS, if a numeric value 3 is typed by replacing the placeholder letter E (the default value text of the VCVS component), it represents a multiplier (or scale factor). A scale factor of 3 implies that a voltage across the output terminals of the VCVS is 3 times the voltage difference between the two control input line terminals. From the layout or orientation of e2, it can be said that its output holds a negative sign (opposite to that of input). The scale factor can also be entered as an equation relating to other circuit parameters.

To change a scale factor of the e2, right-click on the placeholder letter E to open the **Enter new Value for E1** box where the user can enter a desired scale factor as shown in Fig. 4.81. Alternatively, the scale factor can also be changed by pressing Ctrl + right-click (or simply right-click) on the VCVS component body. This action opens the Component Attributes Editor box. The user can edit the scale factor in the text box determined by the Value row, Value column.

ii. Now, place the resistor of resistance 1 kiloohm across the VCVS.

- Connecting an Independent source to the Dependent Source (without wiring)

iii. As we know that in LTspice XVII, all the common nodes or points share a same reference number by default, the components can be interconnected by assigning the same node names or net labels (using the **Label Net** tool) to their respective terminals which are needed to be connected together as shown in Fig. 4.82. Thus, nodes with the same name are actually shorted together.

Therefore, add the net label Node1 to a positive terminal of the V1 and the net label Node2 to its negative terminal. Also, assign the net label Node1 to a - ve control input of the VCVS E1 and the net label Node2 to its + ve control input. Edit the VCVS (symbol e2) scale factor value E to 3.

Remember: The nodes with the same label act as if they are wired together. This feature is used to make complicated schematics look much neater. Thus, the nodes (existing in two separate circuits) with a same label act as if they are interconnected despite the fact that they are not having any visible connection.

(**Note:** An error often arises when we duplicate the circuit (or portion of the circuit) that contains labeled nodes.)

- **Run a DC Operating Point Simulation**

iv. Enter a multiplier (or scale factor) of 3 in the Value row and Value column text box replacing a letter E (see Fig. 4.81). After completing the circuit schematic, run the .op simulation to observe the following results:

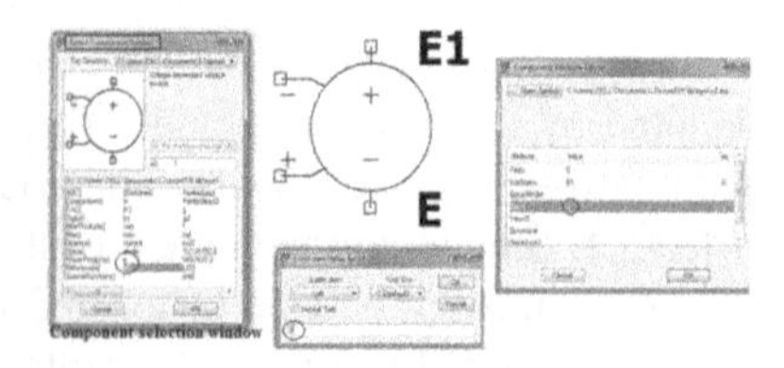

FIGURE 4.81 Editing the Multiplier Factor of the VCVS.

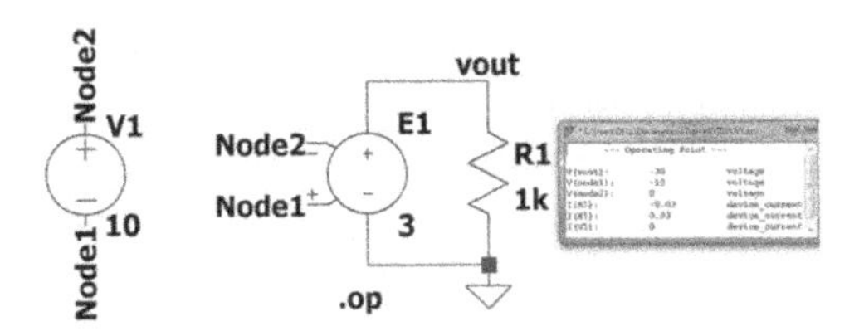

FIGURE 4.82 The Circuit Schematic with Connections without Wiring.

The simulation results show that voltage at the node vout is -30, that is, it is equal to 3 multiplied by 10 volts (or three times the value of the V1) and having a negative sign because a + ve terminal of the V1 is connected to a – ve control input terminal of the E1. A current through the R1, that is, I(R1) is negative because it is leaving out of the R1's default positive terminal. Also, a current through the E1 is positive because it is entering into the source Node2.

v. Similarly, again select the Component icon and choose the symbol **e** to place it on the schematic. Draw the circuit diagram as shown earlier but replacing the component with a symbol **e2** by the component with a symbol **e**. Enter the V1 value to 10 (DC amplitude in volts). Add the net label Node1 to the positive terminal of the V1 and the net label Node2 to its negative terminal. Also, assign the net label Node1 to the +ve control input of the VCVS E1 and the net label Node2 to its - ve control input. Now, run the .op simulation to obtain the following results. Add the net label Vout to the positive output side of the E1. Assign the GND node to the negative output side of the E1. Edit the VCVS value (scale factor) E to 10. The simulation results show that voltage at the node Vout is now positive and equal to 50 V, that is, 10 times the value of the V1.

(**Note:** A size of the Operating Point floating window can be reduced by placing a cursor over the lower right corner, and when a double arrow appears, left-click and drag the window to an appropriate size while holding a mouse button.)

- **Usefulness of the Net Labeling:**

To conclude, labeling the circuit nodes by pressing an F4 key or the Label Net button (a box with an A inside it) can serve the following purposes:

- **Provides logical names and easy identification:** More appropriate names like out and in can be assigned to the nodes to make it easier for the user to identify them in the simulation results. Also, the users can easily pick out the one they want to plot from a list of visible traces by clicking on the schematic sub-menu option Visible Traces that can be accessed by selecting the View schematic menu.
- **Connects without wiring:** If a certain node is connected to so many points in the circuit and wiring them results in a messy drawing, this can be eliminated

by giving all the nodes the same name instead of connecting with wires. For example, label/name a positive node terminal of the voltage source V1 as the Vin, and then put the same netlabel Vin on all the points needed to be connected to the V1 positive terminal. It creates the same effect as is generated when the nodes are connected with wires.

5 Transient Simulations

5.1 INTRODUCTION

Simulating the circuit schematic using a transient analysis allows the user to examine the circuit's behavior (taking a dynamic relationship for each component in the circuit into an effect, that is, the capacitor models are not open-circuited) over time by solving algebraic-differential equations describing it and display the simulation results comparable to an oscilloscope display and measurements in a real circuit. By default, automatically an algebraic or a differential equation is solved numerically using the DC operating point analysis values as the initial conditions. But, the .ic directive (syntax: .ic [V(<n1>) = <voltage>] [I(<element>) = <current>]) can also be used to initialize node voltages and branch currents before a transient analysis is run. In the syntax, V(<n1>) refers to a voltage at the <n1> (angle brackets indicate the parameter that needs to be settled). More than one node voltages/branch currents or each can be specified in one directive by specifying node voltages and branch currents one by one. For example, to specify an initial charge on the capacitor in a discharging capacitor, write .ic V(N002) = 1 to initialize a voltage at the node N002 to 1 V.

In transient simulations, all the time-independent sources are set to DC values and steady-state waveforms in addition to transients are observed to determine a response of non-linear circuits. The drafted circuit response in a time-domain can be examined by adding a sinusoidal source in the schematic circuit. If a driving voltage is composed of multiple AC signals in the circuits having active and non-linear elements, sidebands through frequency mixing are generated which may affect output from the circuit. Therefore, it becomes important to see how a circuit changes the shape of an input signal by observing an output in a time-domain. A transient simulation displays how a circuit affects the shape of actual waveforms and produces distortion.

Transient analysis typically involves using an oscilloscope to observe waveforms in time-domain. Thus, it is the most powerful (realistic) circuit simulation method that allows representing different wave-shapes, modeling non-linear devices, and observing the output of various circuits such as clippers, clampers, power supplies, switching circuits, etc. for arbitrary inputs. In the circuits having active and non-linear devices/elements, the shape of the input signal changes at the output and to obtain an accurate view of how the shape of the signal changes (when the circuits are operated at different input frequencies or amplitudes), it is needed to look at the signal in the time domain.

LTspice XVII allows the independent voltage/current source component to be configured in many possible ways so that by first clicking the Advanced button in the component attribute window, and then changing the Functions setting in the source component menu window, several different waveforms can be generated such as sinusoidal, pulse, exponential, piece-wise linear, etc. Using the PWL function for the independent voltage/current source, a suitable waveform data can be described as a set of any number of points entered directly into the **Time/Value pairs** list, that is, a time position followed by an amplitude. Selecting (by checking the selection box) the SFFM option in the Functions setting of the independent source component menu, generates a single frequency FM modulated sine wave signal. For the current sources, always point an arrow in a direction of the conventionally flowing current.

For generating a sine wave to run a transient analysis, the parameters to be entered for the selected time-dependent SINE function are:

i. **DC Offset:** It is the DC offset voltage that represents how many volts above or below a ground the signal is. It should be set to zero for a pure sinusoid to make the positive and negative peaks equidistant from zero.
ii. **Amplitude:** It is an un-damped amplitude of a sinusoid (peak value measured from a zero level (i.e., with no DC offset value).
iii. **Freq:** It is the frequency in Hertz.
iv. **Tdelay:** It is the time delay in sec (seconds). It should be set zero for a normal sinusoid.
v. **Theta:** It is the damping factor (not a phase angle). Also, set it to zero for a normal sinusoid. Its value is used to apply an exponential decay to a sinusoid (the Theta is a reciprocal of the decay time constant with unit 1/sec or 1/seconds).
vi. **Phi:** It is the phase advance of a sine wave in degrees. Set this to 90° if a cosine waveform is required.
    Note that the phase angle if left unspecified is set by default to 0°
vii. **Ncycles:** It represents a number of cycles of a wave or pulse (neglect for free-running pulse function) that the user wants to happen (leave it to have ongoing pulses).

Thus, during a transient analysis, DC components of the selected source function are entered.

Also, the default time display on an x-axis in a transient analysis can be changed to show other quantities (like current) on a horizontal axis to validate the model parameters.

To change the default time display settings of an x-axis:

- After plotting a voltage/current trace in the waveform viewer, move a cursor to a horizontal axis (or over an x-axis) of the waveform viewer and left-click when a mouse cursor turns into a ruler.
- In the Horizontal Axis dialog window (see Fig. 5.1), enter an expression for the quantity needed to be displayed on an x-axis into the Quantity Plotted field.
- Click the OK.

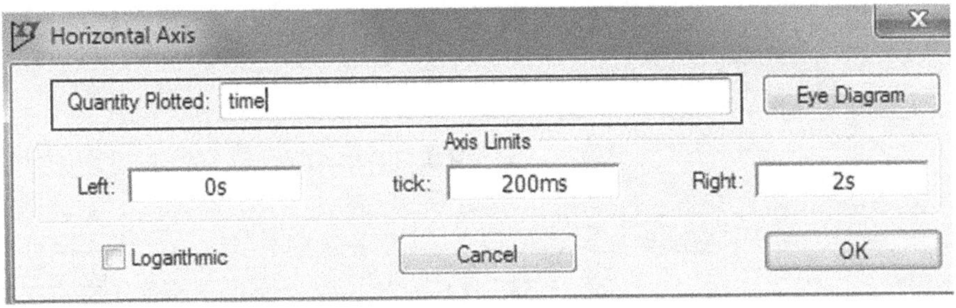

Horizontal Axis
Quantity Plotted: time
Eye Diagram
Axis Limits
Left: 0s tick: 200ms Right: 2s
Logarithmic
Cancel
OK

FIGURE 5.1 Editing the Horizontal Axis Parameters.

The Horizontal Axis dialog window in case of a transient simulation for entering an electrical quantity needed to be displayed on an x-axis into the Quantity Plotted field.

5.2 SINE WAVE GENERATION

Below is the circuit schematic to generate a sine wave from the voltage source component V1 (here, the source positive terminal is connected to an output node using net labeling). Construct the following circuit schematic using the following steps:

To perform a transient analysis, the user needs to set up the independent sources into alternating nature, for example, a source should be initialized to output a sinusoidal waveform.

i. **Create a sinusoidal voltage source as an input signal:** Right-click on the body of the component voltage with the default label V1 to open the source configuration window with standard parameters as shown in Fig. 5.3.

Now, click the **Advanced** (see Fig. 5.3) button to open the voltage source component menu window with advanced parameters as shown in Fig. 5.4. A series of radio buttons on the left side of the window allows the user to choose a desired source wave shape.

The Independent Voltage Source – V1 dialog box (divided into two panes) appears on the screen that allows the user to specify additional properties of the source (shape of a wave) for doing different analysis types. From a list of sources

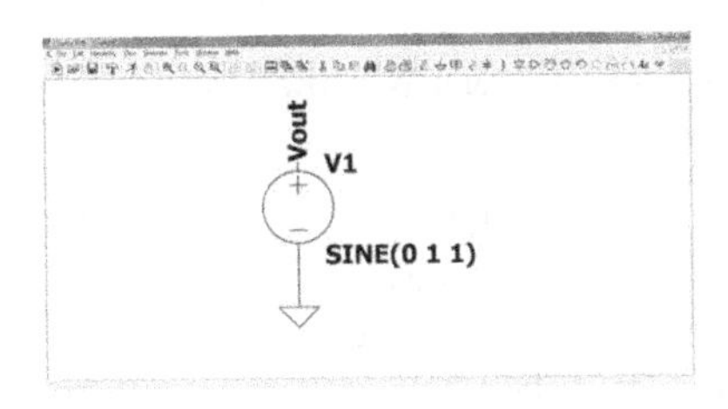

FIGURE 5.2 The Circuit Schematic for a Sine Wave Generation.

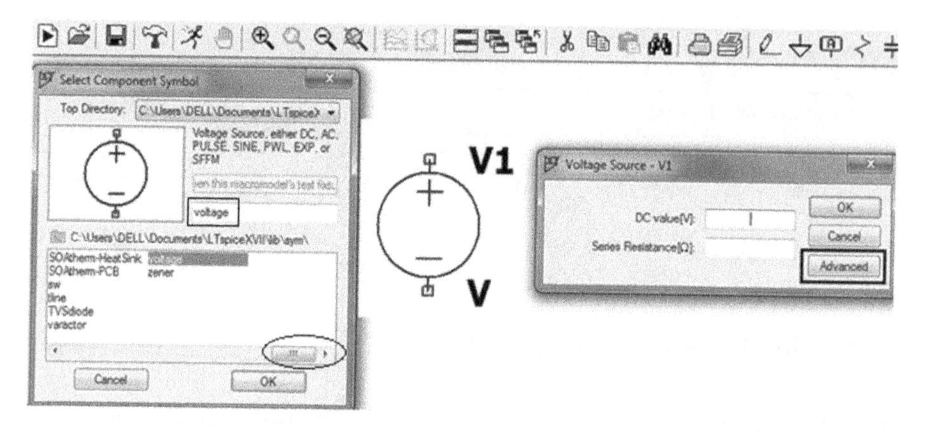

FIGURE 5.3 The Voltage Source Component and its Value Editor Window.

Independent Voltage Source - V1

Functions

(none)
PULSE(V1 V2 Tdelay Trise Tfall Ton Period Ncycles)
SINE(Voffset Vamp Freq Td Theta Phi Ncycles)
EXP(V1 V2 Td1 Tau1 Td2 Tau2)
SFFM(Voff Vamp Fcar MDI Fsig)
PWL(t1 v1 t2 v2...)
PWL FILE: Browse

DC offset[V]: 0
Amplitude[V]: 1
Hz, the default unit — Freq[Hz]: 1
Tdelay[s]:
Theta[1/s]:
Phi[deg]:
Ncycles:

Additional PWL Points
Make this information visible on schematic:
Un-tick to hide entered information

DC Value
DC value:
Make this information visible on schematic:
Small signal AC analysis(.AC)
AC Amplitude:
AC Phase:
Make this information visible on schematic:
Parasitic Properties
Series Resistance[Ω]:
Parallel Capacitance[F]:
Make this information visible on schematic:
Cancel OK

FIGURE 5.4 The Voltage Source Menu Window with Advanced Parametric Options.

under the **Functions** setting on the left pane, select the SINE radio button to set up a sinusoidal source. The target is to create a sinusoid signal of 1 V peak amplitude at a frequency of 1 Hz (hertz). Therefore, in the stack of boxes at the bottom of the same left-pane enter in 0 in the **DC offset** field, 1 (a peak voltage of 1 V, no need to enter the default unit V) in the **Amplitude** field, and 1 (the default unit of a frequency is

Hz, see Fig. 5.4) in the **Freq** field. The user can leave the other fields empty. Click the OK. The **AC Amplitude**, and **AC Phase** on the right side of the window are ignored in a transient analysis and used in an AC Analysis.

(**Note:** Left-clicking the **Advanced** button allows giving the source values for other simulation types, for example, waveform types under the Functions button are for Transient (time-domain) simulations, DC Value is for operating point, DC sweep and transient simulations, whereas Small signal AC analysis(.AC) is for frequency sweep AC simulations)

Remember: The field phi is set to 90 (the default unit of a phase is degree) for a cosine wave. The field Ncycles represents a number of cycles. For the given case, if the parameter DC offset is set to be 4, this creates a sinusoid ranging from 3 V to 5 V and centered at 4 V)

IMPORTANT: To plot a sine wave of angular frequency w = 1000 rad/sec, enter the frequency parameter value to be equal to 1000/(2*pi) in the Freq field box. As the period of a sine wave is taken to be 6.3 ms, therefore setting the Stop Time = 63 ms runs the simulation for 10 cycles of the waveform.

ii. Add the net label Vout to the output node (a positive side of the voltage source component) using the **Label Net** schematic tool. Alternatively, the net names can be added by right-clicking on a wire and selecting the Label Net option (can also highlight the net label using the Highlight Net option) or use a shortcut F4 (click the Edit schematic menu to read other shortcuts) for the net labeling.

iii. Connect the GND node to a negative terminal of the voltage source.
Remember: After drawing the circuit schematic, it is important to label/name nodes in the circuit so that in the simulation results, the user can identify the nodes quickly. Also, labeling the organized schematic circuit becomes practical and all the nodes which are assigned the same name get **shorted** together although **no wire** is placed. This works well with large circuits and for connecting voltage sources to different nodes.

iv. Now, go back to the schematic menu and choose the Simulate -> Edit Simulation Cmd options. Click the **Transient** tab to bring the Edit Simulation Command dialog window. By default, a transient analysis starts at a time equal to zero seconds and goes up to the user-defined final time. However, if an initial time for starting saving a data is specified, the circuit is analyzed so that the waveform data (or output) between zero and the time specified in the Time to start saving data field is computed but not saved and discarded.

A maximum allowed time step parameter or a plotting increment (a suggested value for integrating the circuit equations) can be left unspecified. But, sometimes the simulator may choose a time step too long or too short. A very small plotting increment (timestep) may cause the traced graph to appear cluttered because of the presence of unnecessary points and takes a large amount of simulation time. On the other hand, a very long plotting increment may omit important occurrences over very short periods of time in a graphical output and outputs erroneous results. Therefore, a time step value should be entered in the Maximum Timestep field text box which works best for the drafted circuit.

The simulator may adjust a time step dynamically in any case but cannot exceed it from a value specified in the Maximum Timestep field. A ratio of values entered in the Stop time and Maximum Timestep boxes determine the total number of calculations needed by the simulator to plot a waveform. Generally, a time step should be $1/10^{th}$ or $1/100^{th}$ of a stop time. Also, the user can choose to save a few periods towards the end instead of saving the entire data-set by specifying the Time to start saving data field.

In the **Stop time** field text box, set an appropriate time depending on input frequencies in the circuit. By default, a stop/final time (duration of the simulation) is entered in units of seconds. When a transient analysis is performed, it is must to specify a number of signal periods (that should be covered by a simulation) in the Stop time field. Here, five periods are selected and this corresponds to 5s (5*T, where the T = 1/F is a period of a wave).

- **Set the Transient Analysis Parameters:**

In the Edit Simulation Command dialog window (see Fig. 5.5), enter a stop time of 5 seconds (which implies the signal can be seen for 5 periods) in the Stop time text box, set the Time to start saving data box to 1, and the Maximum Timestep box to 0.05 (1/100th of a stop time). Click the OK.

(**Note:** If a frequency of a waveform is 1 kHz, one complete cycle of the waveform takes 1ms (period of the wave). The transient command should be set to stop a recent simulation at 3ms to cover three periods)

Now, drop in the SPICE directive statement (the dot command) at an appropriate place as shown in Fig. 5.5 for performing a transient analysis:

v. Save the circuit schematic file as sinewave (change the default file name Draft29 to sinewave).
vi. Run the simulation by pressing the **Run** button on the schematic toolbar. A blank waveform viewer window appears. Now, left-click with the voltage probe on the node Vout (or a wire connecting a + ve side of the V1 to the

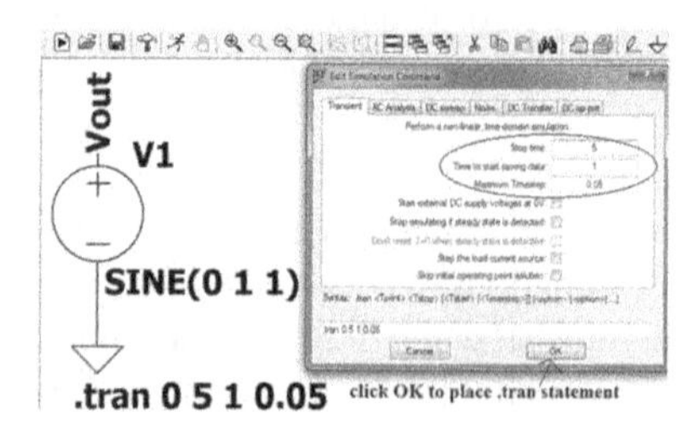

FIGURE 5.5 Editing a Transient Simulation for Generating a Sine Wave.

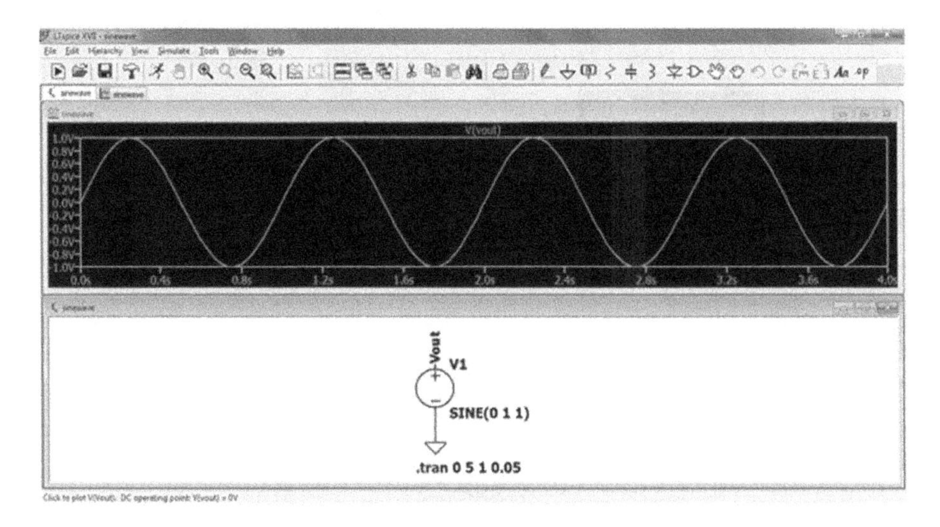

FIGURE 5.6 A Sine Wave Plot in the Waveform Viewer Window.

node) so that an output sinusoidal voltage at the respective node can be plotted with the following default settings:

Thus, a transient analysis SPICE simulation computes a transient output variable as a function of time over a user-specified time interval.

vii. Change the waveform viewer background color to white by going to the schematic menu bar and clicking the **Tools** Tools -> **Color Preferences** Color Preferences schematic menu options. The action brings the **Color Palette Editor** Window. In the **Color Palette Editor** dialog window, select the **WaveForm** tab. Under the **WaveForm** tab, select the **Background** from a scroll-down menu next to the **Selected Item** field, and then move the Red, Green, and Blue sliders to the extreme right (at 255 RGB value) for changing the trace background to white as shown in Fig. 5.7.

Click the Apply and OK to insert the change made (see Fig. 5.8).

- **Editing an Axis of the Trace:**

viii. For the already plotted trace (or to be plotted), the user can change the axis color to black (default is grey) for better readability. Click on the **Control Panel** schematic tool, and then select the **Waveforms** tab from the **Control Panel** dialog window to do the changes as shown in the following:

In the **Waveforms** tab, select the **Color Scheme** button. Now, click on the **WaveForm** tab in the **Color Palette Editor** dialog window, then next to the **Selected Item** box choose **Axis** from the drop-down menu

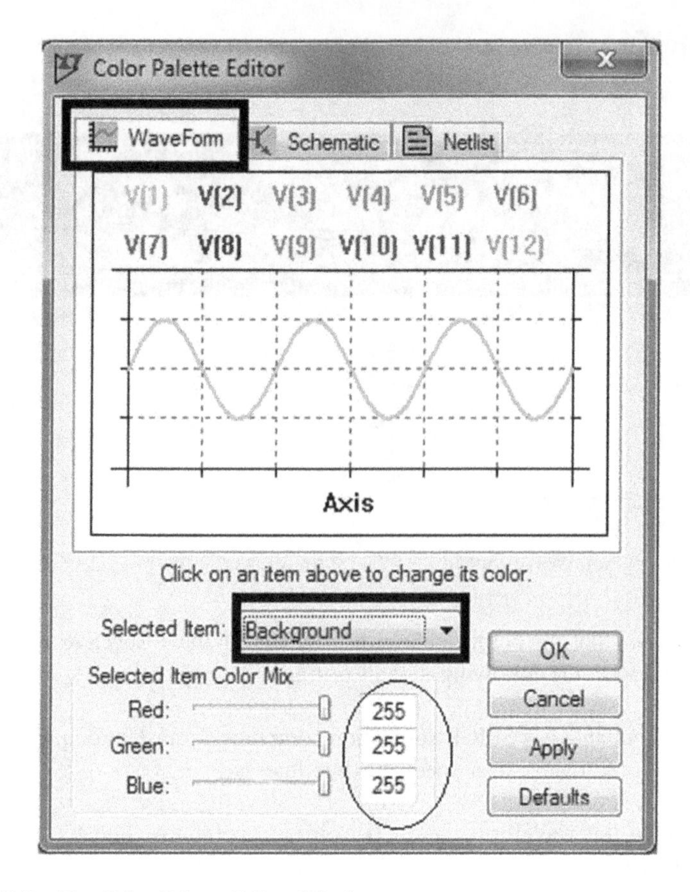

FIGURE 5.7 The Color Palette Editor Window.

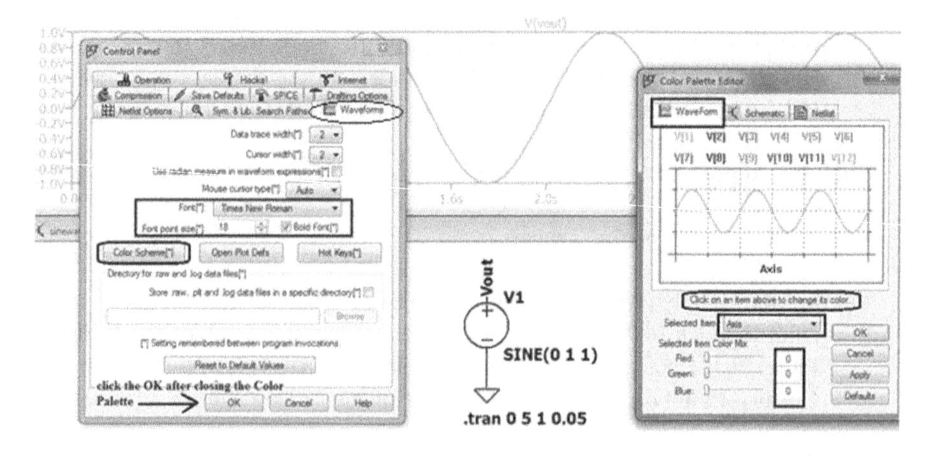

FIGURE 5.8 Editing the Axis Properties of the Plot Pane.

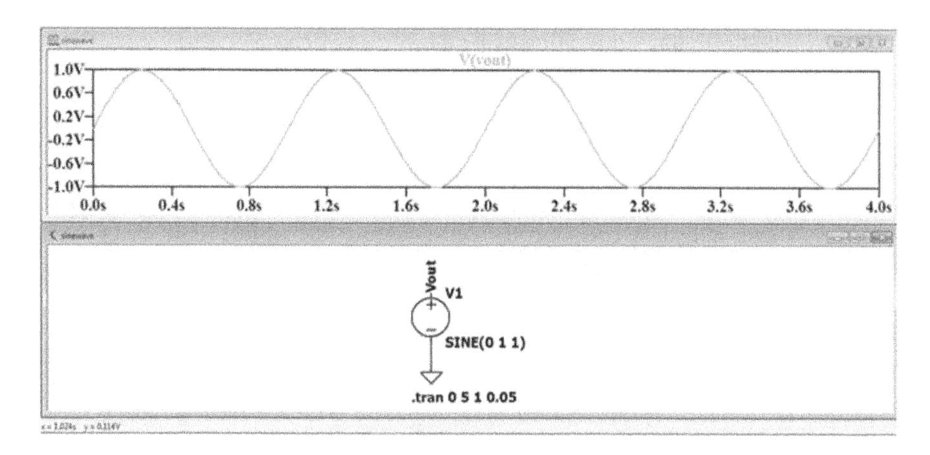

FIGURE 5.9 The Desired Output Sine Wave.

Selected Item: Axis. Then, move the Red, Green, and Blue **sliders** to the leftmost position (i.e., to 0 RGB value) Selected Item Color Mix Red: 0 Green: 0 Blue: 0 to pick a black color for the axis of the plotted graph. Click the Apply and OK. Now in the Control Panel, under the Waveforms tab Waveforms, set a font type to **Times New Roman** and size to **bold** and **18** for editing a text on the trace. Click the OK. The following results are obtained.

(**Note:** There is a mouse cursor readout or status bar called the Information Bar. As a mouse is moved over the traced plot in the waveform viewer window, the mouse position can be read on this status bar. If a mouse is dragged over the wave, the size of a selected box (dx and dy) is displayed on the Information Bar)

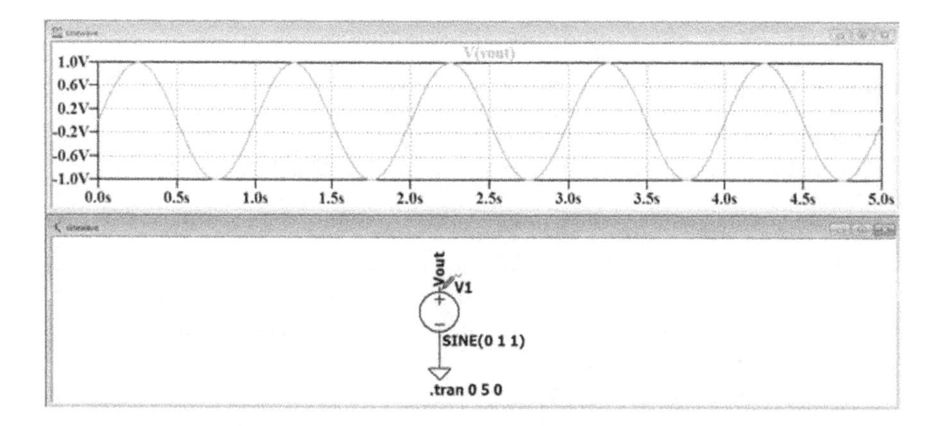

FIGURE 5.10 The Output Sine Wave for a Simulation Time of 5 s.

ix. Probing the output for plotting using the voltage probe cursor when the **Time to start saving** data box is set to 0 and the **Maximum Timestep** field is kept unspecified:

The wave starts at zero seconds and goes up to the user-defined final time, that is, 5 s. A grid is displayed in the given output waveform window by selecting the **View** schematic menu and then checking (or right-ticking) the **Show Grid** check box Show Grid Ctrl+G or pressing Ctrl + G.

(**Note:** By default, LTspice XVII always starts a wave at zero seconds which goes up to the user-defined final time)

Remember: Before copying a plot to the clipboard so that it can be added to a document, it is always better to change the default black background to white.

5.2.1 Graphical Measurements

- RMS Value using the .meas:

The electrical quantities for the simulated graphical output can be easily measured using the .meas directive statement. The .meas directive is a multipurpose directive for taking measurements on the simulation results in a graphical form such as computing the maximum/peak and minimum values of a waveform, finding out when a rising/falling voltage reaches a particular value, calculating root mean square (RMS) currents/voltages, PP (peak-to-peak) and AVG (average) values of a trace, finding INTEG (integration), etc.

The .meas directive command is used to analyze an output data of the .ac, .dc, .op, .tf, .tran, or .noise simulation. The command can be executed after running the desired simulation and is entered in the schematic in the same way as other SPICE directives

To conclude, after completing a simulation analysis, LTspice XVII .meas function allows:

Measuring a y-axis value of a particular point **at** a desired x-axis value or **when** a condition is met,

Finding a maximum/minimum value on a curve,
Computing a peak-to-peak value or an RMS/AVG value,
Measuring a range on an x-axis between two particular points, etc.

x. The target is to measure an RMS value for the sine wave generated with the transient simulation parameters **Stop Time** set to 5, **Time to start saving data,** and **Maximum Timestep** left unspecified. Experimentally, a digital multimeter measures RMS values of AC voltages and currents. But in LTspice XVII, an RMS value can be extracted by using the .meas directive statement. Measuring an RMS value using the .meas statement refers to a

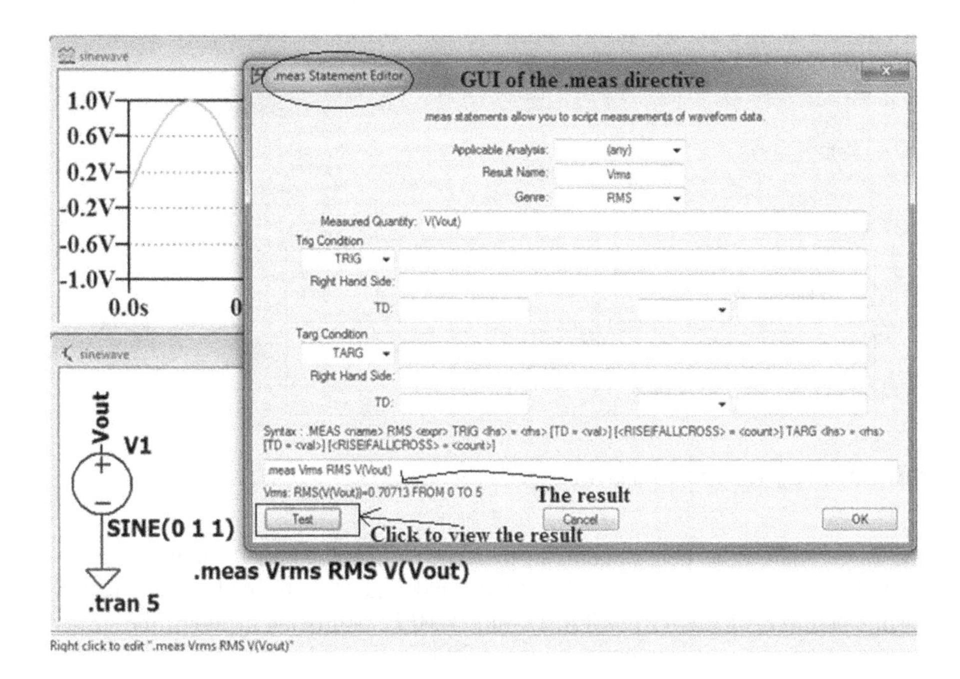

FIGURE 5.11 The Test Result of the .meas Statement.

range over an abscissa or independent variable plotted along a horizontal or x-axis (time axis in the .tran analysis or frequency axis in the .ac analysis).

To place the .meas simulator directive (or any other) on the circuit schematic shown in Fig. 5.11, click on the SPICE Directive schematic toolbar (the .op icon). A text box appears in the Edit Text on the Schematic window where the user can type the directive, say .meas for now. After pressing the OK, the user needs only to place .meas attached to a cursor anywhere on the schematic. LTspice XVII provides a convenient GUI to edit or set most of the available directives. To set the directive statement by accessing the built-in GUI, press a right mouse button on the placed .meas directive. The built-in GUI for the .meas directive looks like as shown in Fig 5.11.

Click on the drop-down menu in the Genre (having many operations to be performed over a range of abscissae such as AVG, PP, MAX, MIN, and RMS) field, select the **RMS** option. To measure an RMS value the generated sinusoidal AC voltage across the specified node enter V followed by the default/user-defined node number or node name. Click the OK and the complete statement gets placed automatically on the schematic.

Now, right-click on the already placed .meas complete directive statement to re-open the .meas Statement Editor, and afterward left-click the **Test** button to view the result.

A range over an abscissa is specified with points defined by the TRIG and TARG Condition fields. The TRIG point for the already defined variable time entered in

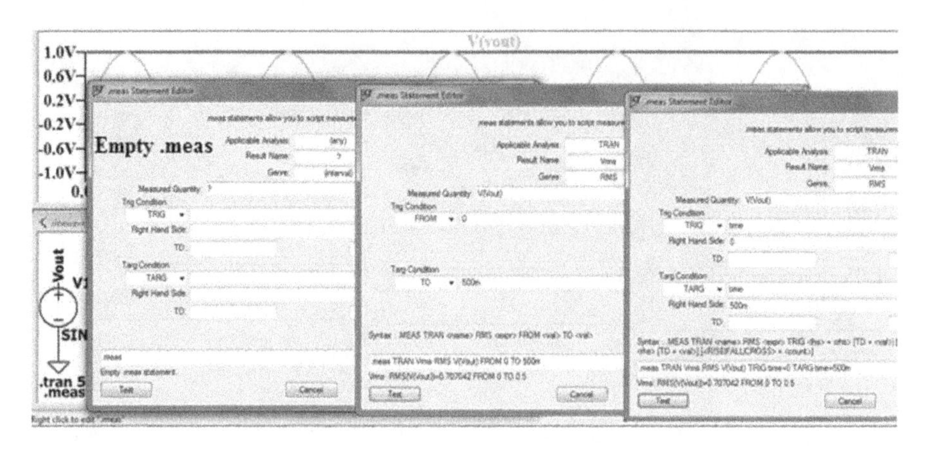

FIGURE 5.12 The RMS Value over a Specified Range.

the Right Hand Side field defaults to the start of a simulation if omitted and the TARG point entered in the Right Hand Side field defaults to the end of a simulation data. Therefore, if the TRIG, TARG, and WHEN points are omitted, the .meas statement operates over an entire range of data. In Fig. 5.11, the TRIG and TARG points are kept unspecified. Thus, an RMS value is computed over an entire range, that is, from 0 to 5 s.

To find an RMS value of the trace from 0 to 500 ms, do the entries as follows (see Fig. 5.12):

Thus, an RMS output value of the trace from 0 to 500 ms can be determined by placing the .meas directive as follows:

.meas TRAN Vrms RMS V(Vout) FROM 0 TO 500 **m**

or

.meas TRAN Vrms RMS V(Vout) TRIG time = **0 TARG time** = 500 **m**

Another example of using the interval .meas statement:

.meas TRAN V AVG V(N001)
+ TRIG V(N002) VAL = 1.5 TD = 1.1u FALL = 1
+ TARG V(N003) VAL = 1.5 TD = 1.1u FALL = 1

Prints an average value of V(N001) from the 1st fall of V(N002) to 1.5 V after 1.1 μs and the 1st fall of V(N003) to 1.5 V after 1.1 μs and label the output as V.

When a measurement operation is not specified, the result of the .meas statement is simply a distance along an abscissa between the TRIG and TARG points.

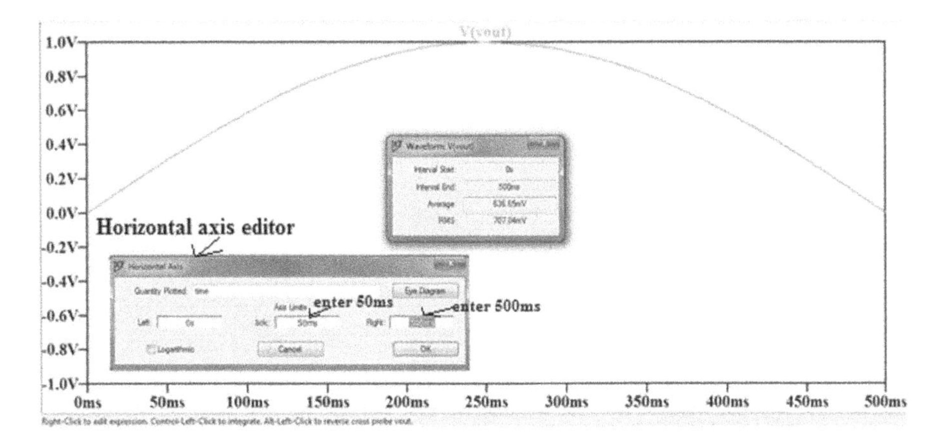

FIGURE 5.13 Editing the Horizontal Axis to Find RMS and AVG Values over a Region.

- **RMS Value using the Axis Editor Window:**

xi. Or else, set an x-axis of the plot to a desired end time, say 5 s. Now, use Ctrl + left-click on the trace label in the waveform viewer window to see the average and RMS numeric values in a little window (not graphs). The floating little window that appears tells the average value (or a definite integral bounded from 0 to 5 s) for 5 cycles of the trace.

But, we can set an x-axis of the traced plot manually to the desired start and end time (by moving a cursor on a horizontal x-axis and right-clicking when a ruler appears) to find a definite integral for a region of interest, say 500 ms equal to a half cycle of the sine wave of frequency 1 Hz as shown in Fig. 5.13.

Move a mouse over a horizontal time axis and right-click when a small ruler appears. This action brings the Horizontal Axis dialog box (see Fig. 5.13). Now, to find the integral over **one half of the waveform Time Period T = 1 sec** set the abscissa from 0 (leftmost value) to 0.5 sec or 500 milliseconds (rightmost value) by changing the entries of the Left, tick, and Right text boxes under the **Axis Limits** field as:

Left: 0 ms,
tick: 50 ms, and
Right: 500 ms

Click the OK to close the axis dialog box and observe the output wave. Now, use Ctrl + left-click on the traced plot label to find the RMS and Average output values in the floating window as shown in Fig. 5.13.

- **RMS Value using the Zoom Tool (dx and dy measurement):**

xii. Alternatively, use the **Zoom to rect**angle tool for zooming the waveform to a region of interest. Thus, select a size of the trace or an

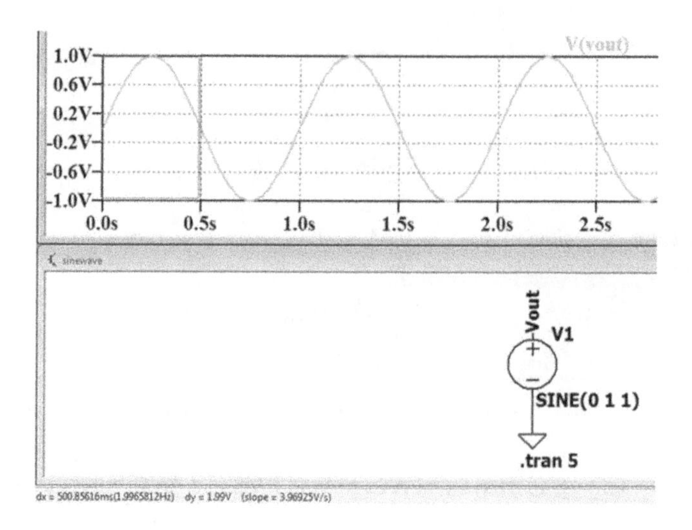

FIGURE 5.14 The dx and dy Value Measurements of the Traced Plot.

interval on an x-axis over which a definite integral is needed to be computed by left clicking over a vertical axis upper limit and drag a mouse up to a desired horizontal axis limit (an abscissa is from 0 to 0.5 s, that is, **one half of the waveform time period T**). After selecting an area and releasing a mouse button, a size of the area can be viewed in the Information Bar at the lower left-hand corner as shown in Fig. 5.14.

Now, use Ctrl + left-click on the traced plot label to find the value of a definite integral bounded from 0 to 0.5 s or average value for one half of a cycle of the trace. To zoom back to a normal area, right-click and select the Zoom to fit or press Ctrl + E.

Thus, transient analysis is a very important analysis which computes various values of the drafted circuit over time and plot a trace against time. In other words, it tells the behavior of a circuit over a specified length of time.

5.2.2 Sinusoidal Signal Source Parameters as Variable/Expression

xiii. A mathematical expression or user-defined variable for the independent source (voltage or current) parameter values can also be entered as well as plotted, but the entire expression must be enclosed in curly braces as shown in the following Fig. 5.15:

The simulated results shown earlier are obtained after running a transient simulation for 20 ms.

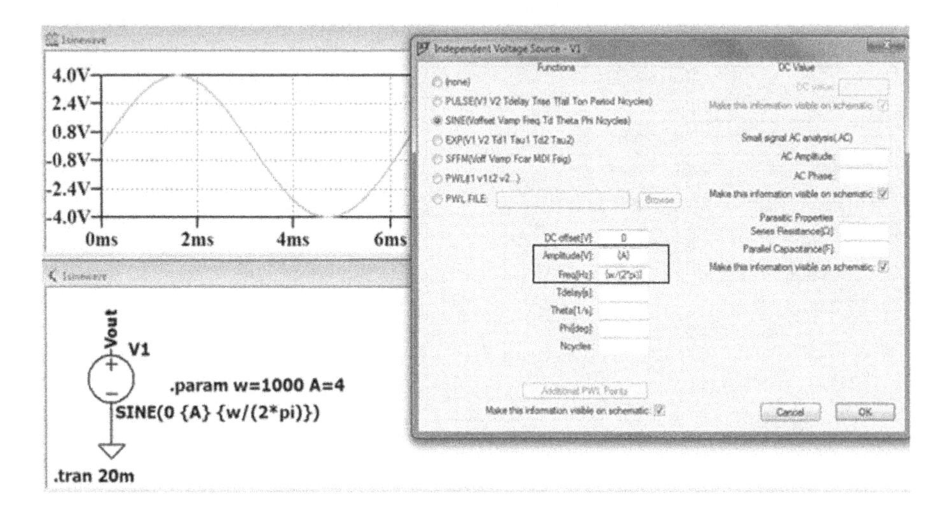

FIGURE 5.15 The Sine Wave (Angular Frequency w = 1000 rad/sec) using the .param.

- **The Sine Wave Amplitude Measurement using the .meas:**

xiv. Now, measure the amplitude value of the generated sine wave with w = 1000 rad/sec at a specific point on time (horizontal) axis using the point type .meas directive as shown in Fig. 5.16:

The point type .meas statement placed on the schematic as shown in Fig. 5.16 returns a value referring to a single point on a horizontal axis. Thus, the .meas statements referring to a single point on a horizontal axis prints a data value or expression at a specific point or when a condition is met. The analysis type is

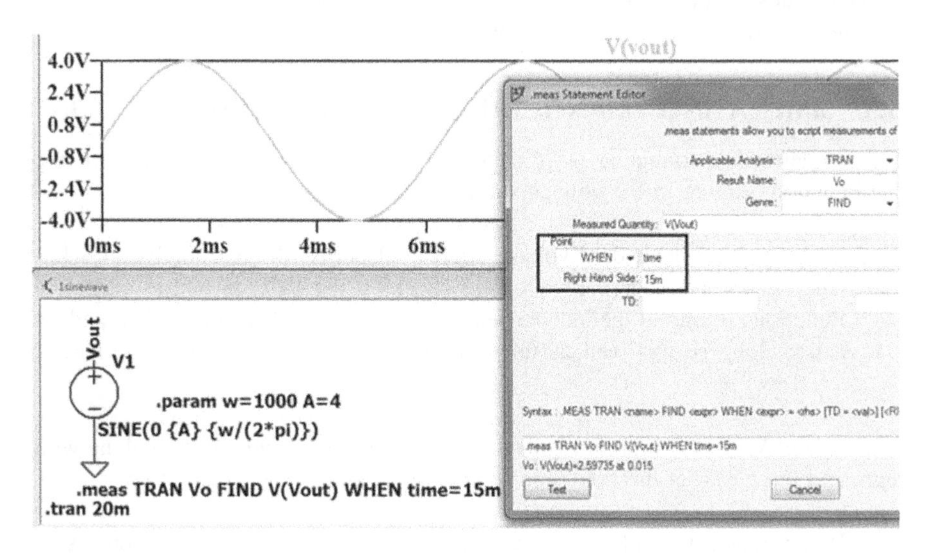

FIGURE 5.16 The Sine Wave Amplitude for a Specified Point in Time.

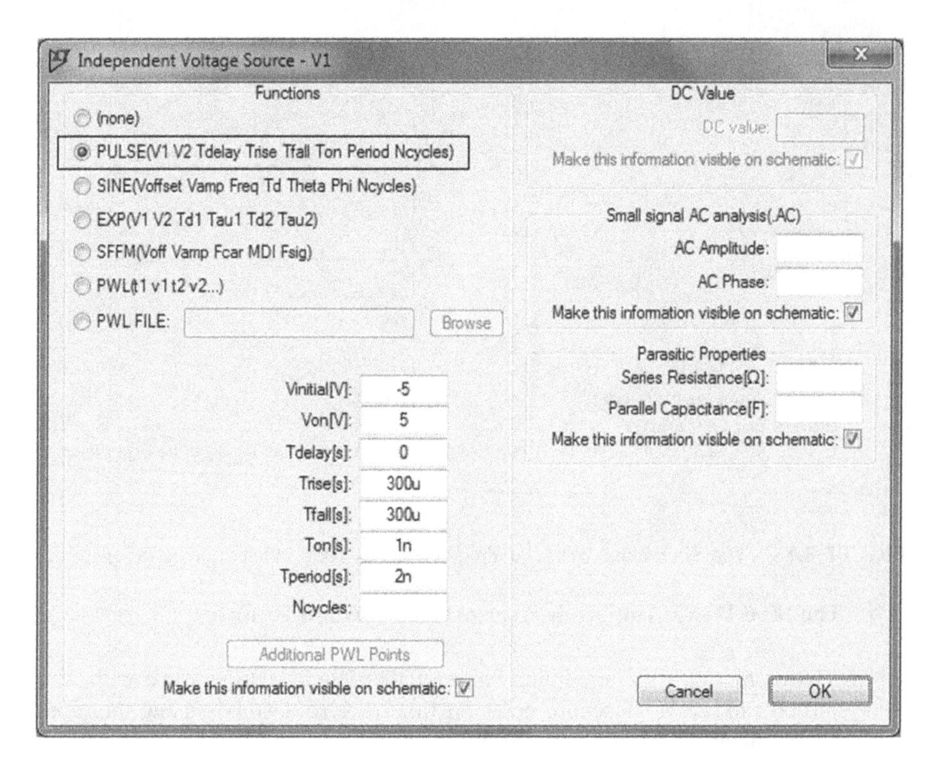

FIGURE 5.17 The Voltage Source Menu Window with Advanced Parametric Options.

generally optional but can always be specified to tell the type of analysis to which the .meas statement is applied. The result of the measurement is contained by a variable name defined in the Result Name field that can be further used as a parameter in the other .meas statements.

5.3 SIMULATING TRIANGULAR WAVE WITH PULSE FUNCTION

How to generate a triangular wave using the PULSE radio button available under the Functions setting in the voltage source component menu window:

i. Connect the independent voltage source V1 across the resistor R1 of 1 kΩ. Choose a pulse source function by checking the PULSE button under the Functions setting in the independent voltage source V1 menu window and set a very long rise as well as fall time, say 300 μs but a short pulse length, say 1 ns.

Set the field Vinitial (minimum voltage value) equal to –5 (enter only 5 along with sign, and not 5V). Set the field Von (maximum voltage value) equal to 5.

i. Run a transient simulation for 5 ms and probe a voltage at the node **out** using the voltage probe to obtain the output as shown in Fig. 5.18:

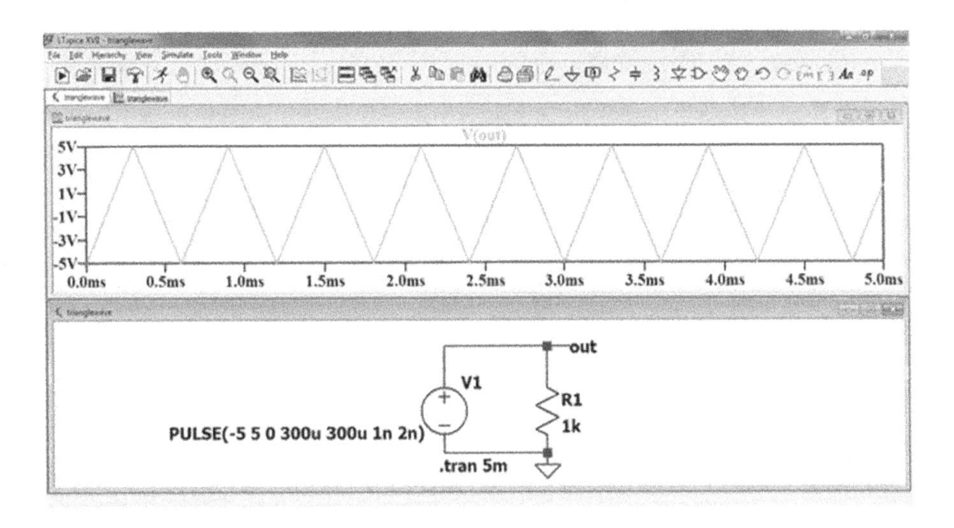

FIGURE 5.18 Triangular Wave Generation.

xv. **Simulating a Square Wave with the PULSE Function**

i. Connect the independent voltage source V1 across the resistor R1 of 1 kΩ. Choose a pulse source function by checking the **PULSE** button under the **Functions** setting in the independent voltage source V1 menu window and set a very short rise as well as fall time, say 1 ns but a quite longer pulse length, say 600 μs.

ii. Run a transient simulation for 5 ms and probe a voltage at the node out using the voltage probe cursor to obtain the square wave output as shown in Fig. 5.20:

5.4 SIMULATING BJT AS AMPLIFIER

i. Draw the circuit diagram of a common-emitter amplifier by picking the NPN transistor model 2N3391A with an emitter resistor taking part in an AC operation.

ii. Select the SINE function for the voltage source and assign numeric values to all the device components as shown in Fig. 5.21.

iii. Select a transient analysis for simulation. Choose the Simulate -> Edit Simulation Cmd options from the schematic menu. Click the **Transient** tab (under the Edit Simulation Command dialog box) and enter 2 in the **Stop time** text box to run a transient simulation for 2 seconds (equal to 10 times the time period of a sine wave having an amplitude of 1 volt and frequency of 5 Hz).

(**Note:** If the circuit diagram is zoomed out or zoomed in, hit a Space Bar or

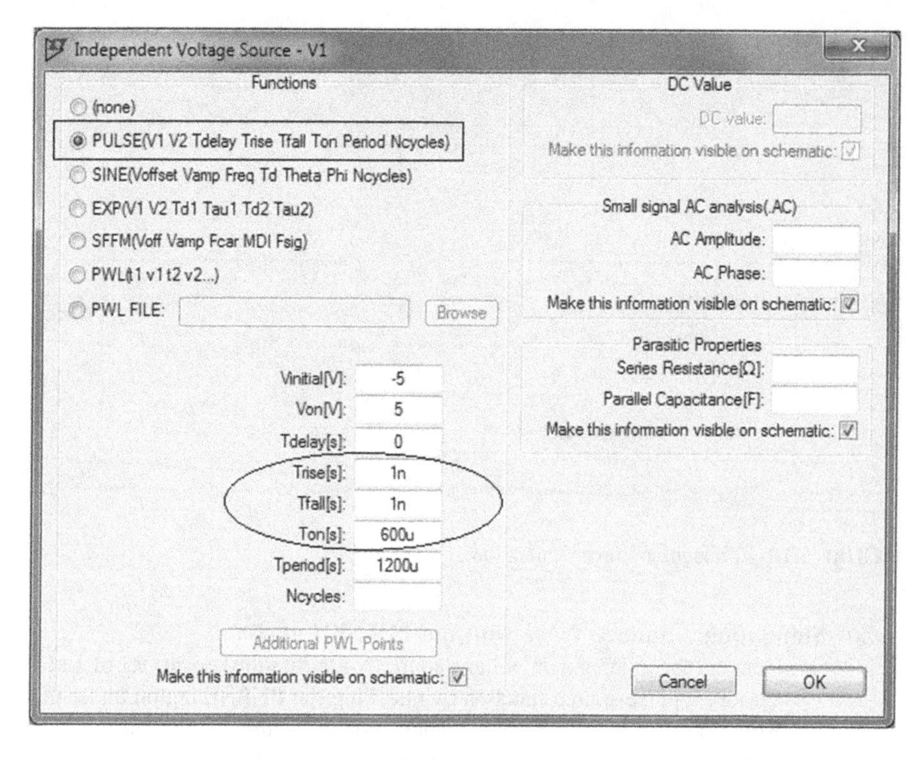

FIGURE 5.19 The Component Menu Window.

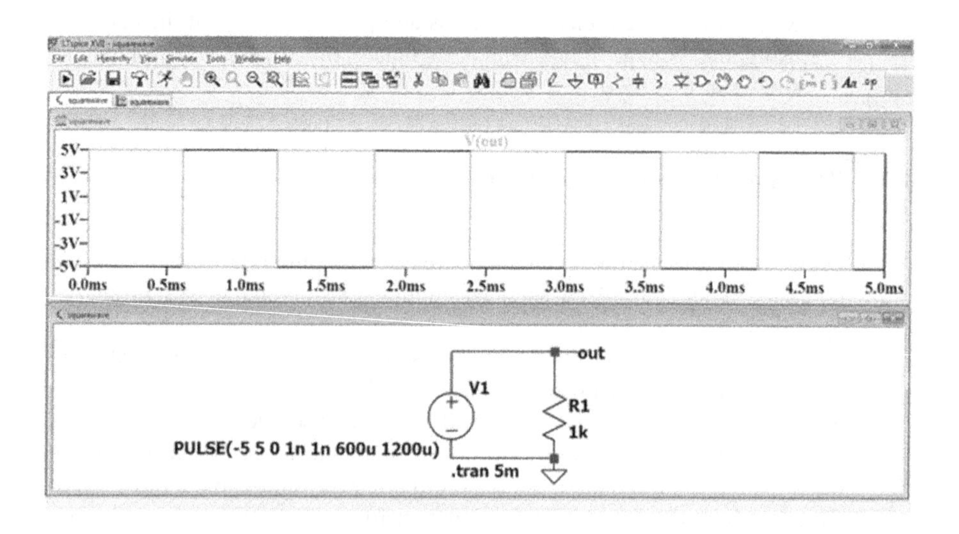

FIGURE 5.20 Square Wave Generation.

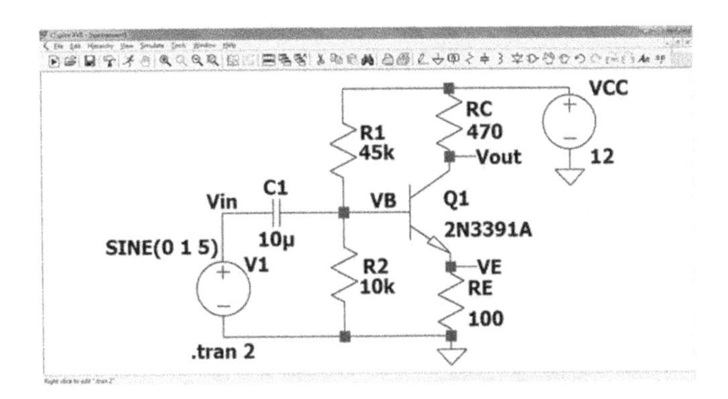

FIGURE 5.21 The Voltage-Divider Biased CE Amplifier with Emitter Resistor.

click the Zoom full extents schematic tool to fit the drafted circuit schematic onto a screen)

iv. Run the simulation by pressing the **Run** button on the schematic toolbar. A blank waveform viewer window appears. Probe the base and collector voltages (click with the voltage probe at the Vout and Vb) and plot them in different window panes. Move a cursor (cross-hair or plus sign) on the node Vb and click a left mouse button (when the voltage cursor appears) for plotting an input voltage at the base of the transistor. Add additional window pane (click on an empty area in the already displayed waveform viewer window to pop-up the plot menu and then go to the Plot Settings -> Add Plot Pane), and afterward left-click with the voltage probe on the node Vout to plot an output voltage at the collector as shown in Fig. 5.22:

The amplified output is 180° out of phase with the input sine signal.

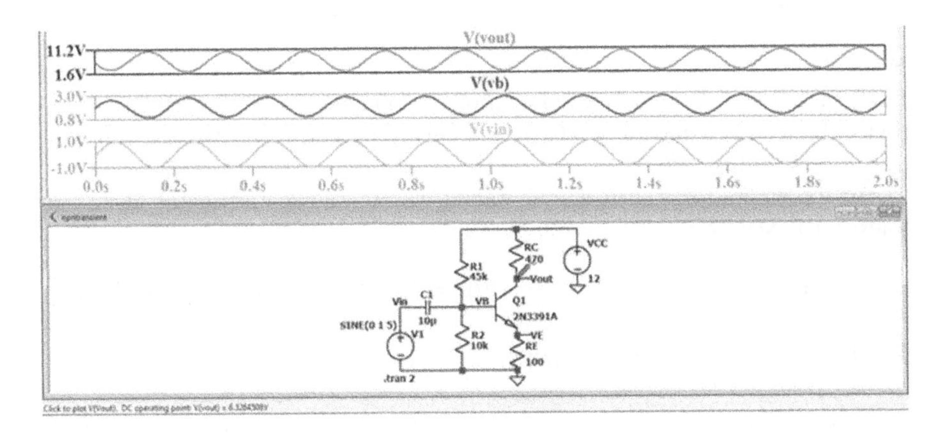

FIGURE 5.22 The Amplified Output.

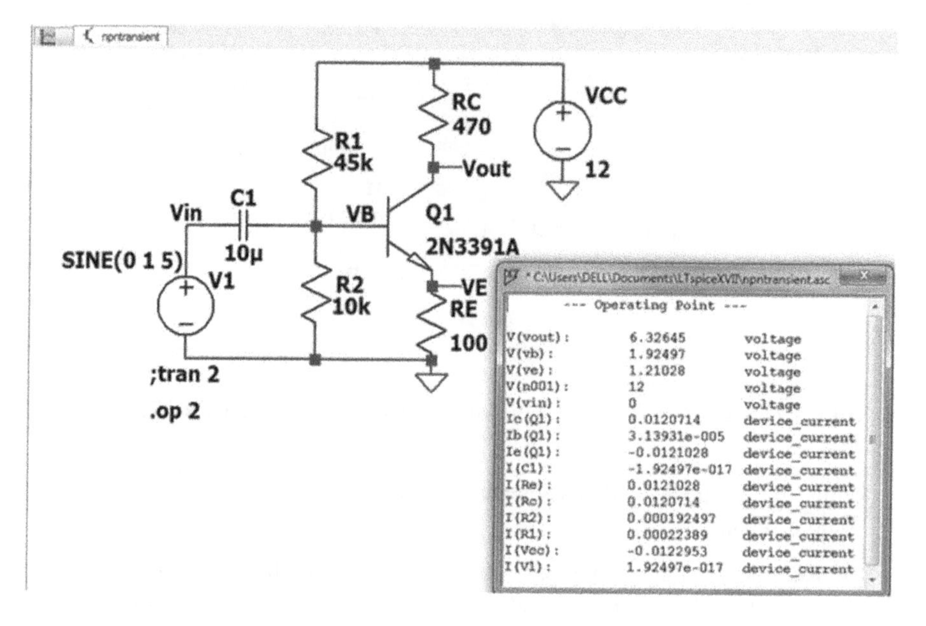

FIGURE 5.23 The.op Simulation of the Designed Transistor Amplifier.

(**Note:** For a multiport device component such as a transistor having more than two nodes, right-click where the component connects to the circuit to probe individual currents)

Choosing Trace Labels from a List in the Visible Traces for Plotting:

Also, the user can pick out another trace for plotting from the visible traces list by selecting the View -> Visible Traces options from the schematic menu.

v. Close the waveform viewer windows opened in different panes. Now, re-run the .op analysis.

From Fig. 5.22, it can be easily stated that the DC levels in the input and output are different and correspond to the solution of the .op analysis of the drafted voltage-divider biased common-emitter (CE) amplifier without the emitter bypass capacitor as can be seen from Fig. 5.23.

For the earlier-mentioned drafted circuit, an AC emitter resistance of an npn transistor model 2N3391A is given by:

AC emitter resistance = $r_e' = 25\ mV/I_E = 25\ mV/0.0121028A = 2.0647$ ohms.

5.4.1 Voltage Gain Measurement

vi. To show that the output variations are amplified, use the LTspice XVII numbered cursors. Enabling both the cursors (**allows automatic math by**

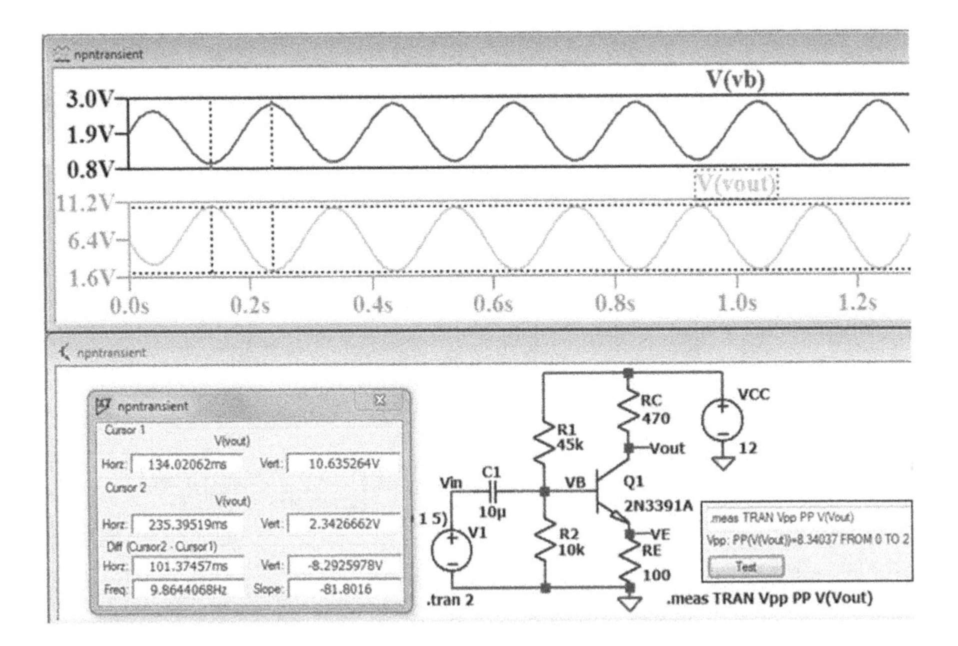

FIGURE 5.24 The Read-Out Display Using both the Numbered Cursors.

giving a difference between them in both dimensions), an amplitude difference between the adjacent negative peak and positive peak (maximum voltage change occurs during one cycle of an AC voltage/current) is measured to find the peak-to-peak output voltage (see Fig. 5.24). While computing a peak-to-peak value, the first peak that is significantly lesser than the rest of the waveform is always ignored. Right-click on the output trace label/name V(vout) to select the 1st & 2nd option from the drop-down menu next to the Attached Cursor field so that two cursors (the Cursor 1 and the Cursor 2) can be attached simultaneously on the same trace. Move the Cursor 1 on the first positive peak and the Cursor 2 on the successive negative peak of the output trace so that the time difference between them can be read around **100** ms (equal to one-half of the complete one cycle period) in the **Horz** field under the Diff (Cursor2 – Cursor1) field as shown in Fig. 5.24.

From the floating cursor window under the Diff (Cursor2 – Cursor1) field, the peak-to-peak output voltage (a difference between vertical position of the Cursor 2 and Cursor 1) can be read as −8.2925978 V.

Also from Fig. 5.24, it can be observed that at a positive peak of the output there occurs a negative peak for the input voltage at the base. Therefore, it can be stated that the output is in phase opposition to the input.

Now, click on the other trace label V(vb) to attach the numbered cursors and measure the amplitude difference between the adjacent negative peak and positive peaks to find the peak-to-peak input voltage.

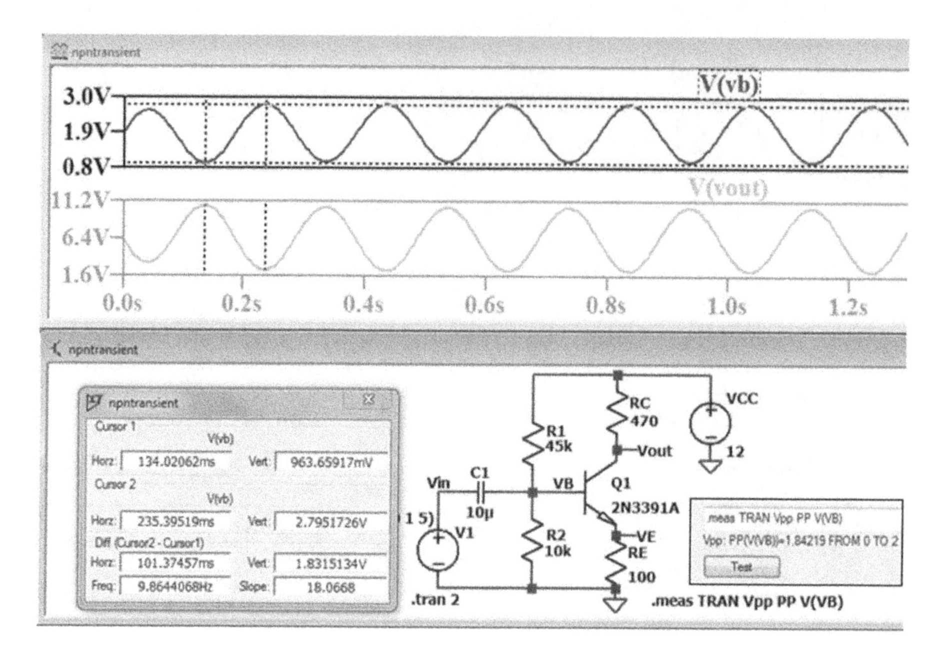

FIGURE 5.25 The Read-Out Display for the Peak-to-Peak Input Voltage.

From the little floating window that pops up, the peak-to-peak input base voltage (difference between the vertical positions of Cursor 2 and Cursor 1) can be read as = 1.8315134V.

Thus, it can be stated that:

- **Peak-to-peak Output Collector Voltage** = – 8.2925978 V
  Otherwise, the peak-to-peak output voltage can be measured by placing the following .meas statement:
- .meas TRAN Vpp PP V(Vout)
  The result comes out to be:
  Vpp: PP(V(Vout))=8.34037 FROM 0 TO 2
- **Peak-to-peak Input Base Voltage** = 1.8315134 V

Similarly, the peak-to-peak input voltage can be measured using the following .meas statement: .meas TRAN Vpp PP V(VB)

Vpp: PP(V(VB))=1.84219 FROM 0 TO 2

Thus, from the given computed values, it can be stated that:

Voltage gain (ratio of collector-to-base voltage) = −8.2925978/1.8315134 = −4.5277297998 = −4.528 (after rounding)

Manually, a voltage gain of the CE amplifier **without** an emitter bypass **capacitor** is given by:

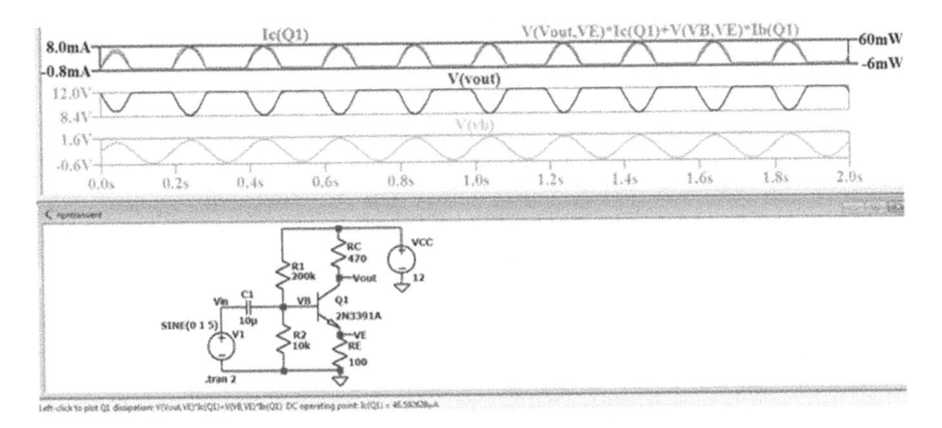

FIGURE 5.26 The Transistor with Non-Linear Amplification (Distorted Output).

$$A_v = \frac{R_C}{r_e' + R_E}$$

$$= -470/(2.0647 + 100) = -4.6049$$

Without a bypass capacitor, an emitter is no longer at AC ground. Instead, R_E is seen by an AC signal between an emitter and ground and effectively adds to r_e' in a voltage gain formula.

Thus, the simulated gain matches the theoretical gain.

5.4.2 Output Distortion (By Shifting an Operating Point)

vii. If the R1 is changed to 200 kiloohms (so that a voltage at the base is reduced and so the emitter voltage), the collector-emitter voltage V_{CE} becomes equal to (11.9781 – 0.0046865 = 11.9734 V) and the operating point shifts too close to the cut-off point, that is, 12 V. As the transistor is driven into cut-off, a positive portion of the output is clipped off as shown in Fig. 5.26.

The output is distorted as amplification is no longer linear.

Also, if the amplitude of the input AC signal is increased in steps while doing the transient analysis, a point is reached where distortion begins to occur and the tops and/or bottoms of the waveform begin to flatten.

IMPORTANT: Probing using Alt + left-click on the component displays a power dissipation of the component (a cursor changes to a thermometer) and with Alt + left-click on the wire displays a plot of the current through the wire (a cursor transforms into a multimeter with an arrow).

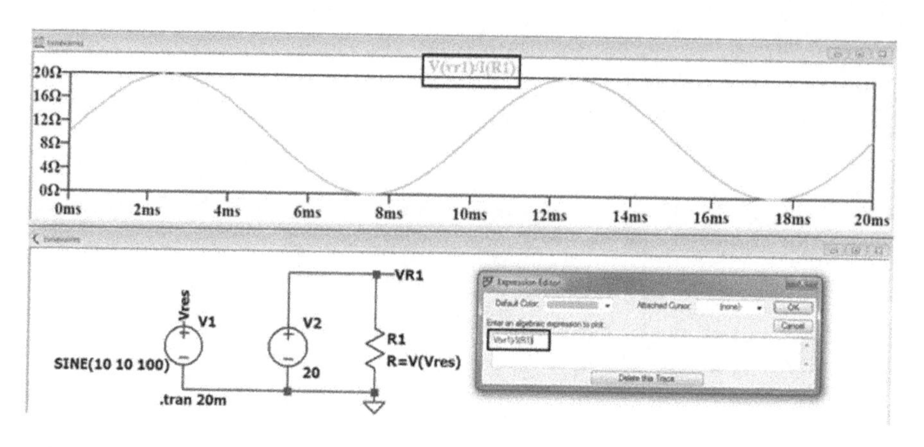

FIGURE 5.27 Resistance Plotted as a Time Variable Parameter.

5.5 RESISTANCE AS FUNCTION OF TIME

- **Simulating the resistor component whose resistance varies in a sinusoidal fashion between 0 Ω and 20 Ω at a frequency of 100 Hz:**
  i. Variable resistance can be obtained from the resistor component by controlling the resistance parameter using the equivalent time variable voltage source, say sinusoidal in this example. For implementing the said action, first place the independent voltage source V1 on the schematic and select the SINE function for setting up the voltage source and give a descriptive name (e.g., Vres) to its output (positive terminal) node using the net labeling. Now, place another voltage source V2 and set it to a DC value of 20 V. Connect the resistor component R1 across the V2. Connect the negative terminal of all the components to a common ground using the Ground symbol as shown in Fig. 5.27. The voltage component representing the time variable behavior can also be set up by selecting any other suitable waveform, Piece-Wise Linear (PWL) function, PWL FILE function, etc.

     Assigning the component parameter value: A constant real number or an algebraic expression of real values, the predefined functions, user-defined functions or the circuit model variables can be assigned to the parameter values of the components.

  ii. Thus, in the value field of the resistor component (whose resistance parameter needs to be simulated as a time variable quantity), set the value to R = V(Vres) so that after the simulation 1 V is equivalent to 1 Ω, 1 kV is equivalent to 1 kΩ, etc. Run a transient simulation for 20 ms. Probe a voltage at the node VR1 and using the Expression Editor (right-click on the trace label V(vr1) plot the trace V(vr1)/I(R1) which corresponds to the resistance of the R1.

The Output:

Alternatively, by setting V(vr1)/I(R1) in the **Expression(s) to add** text box in the **Add Traces to Plot** dialog window (having the list of available data) obtained from the **Add Traces** context-based menu option on the waveform viewer of the simulation result, the resistance can be made variable to a time axis after executing a transient simulation. By changing the resistance value in a transient analysis (time analysis), the voltage and current changes can be observed as the resistance changes.

Now, after closing the waveform viewer window, if the user left-clicks on the VR1 or the vertical wire segment, the voltage value gets displayed.

(**Note:** LTspice XVII is also capable of plotting theoretical expressions alongside simulated results. When the waveform viewer window is visible, the user can move a cursor over a vertical axis or horizontal axis to edit the axis parameters by left-clicking when a cursor changes into a ruler shape. This action brings up the dialog window where the user can change the parameter options corresponding to these axes under various fields)

5.6 ARBITRARY BEHAVIORAL SOURCES

When a source is defined with an **arbitrary expression,** it is called a behavioral source. LTspice XVII provides the arbitrary behavioral voltage/current sources (search for the component symbol name and syntax in the Help -> Help Topics and type sources in the text box under the Search tab). The expressions used for defining a source behavior may contain the node voltages, node voltage differences, circuit element currents, keywords/built-in functions/user-defined functions (like time, pi, temp, delay, if statement, etc.).

i. Search for an arbitrary behavioral current source by typing a character **b** (or bi, bi2, bv) in the **Select Component Symbol** dialog window. The difference between the bi and bi2 is the direction of current flow, whereas the symbol name **bv** represents the arbitrary behavioral voltage source.
ii. Place it on the new schematic and open its attribute editor window:

Enter an expression (containing a valid if statement) in the Value row and Value column field having the syntax as follows:

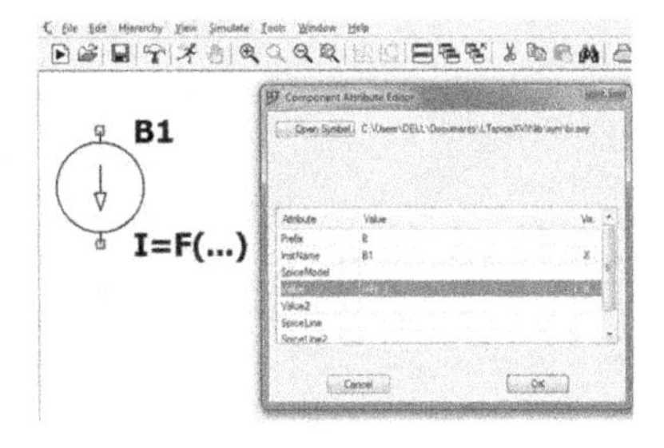

FIGURE 5.28 The Behavioral Current Source Attribute Editor Window.

$$I = 10/(\text{if}(V(\text{node1}) < 5,2,1))$$

iii. Connect the behavioral current source B1 across the independent voltage source component V1.

iv. Now, after placing the independent voltage source V1, open the component menu window by hitting the **Advanced** button in the component attribute editor window. Select the Piece-wise linear voltage/current source by checking the **PWL** radio button under the Functions setting in the component menu window. The PWL function can create arbitrary waveforms by allowing the user to define data points (tz, vz) for drawing a precise waveform, where the t represents time, v represents amplitude value and z is an integer in an increasing order that can be of any value. For times before t1, the voltage is v1. For times between t1 and t2, the voltage varies linearly between v1 and v2. For times after the last time, a voltage follows the last value. For fixed sources, the output (voltage or current) is constant at the end value, but for other types, the last segment is extended linearly that can only be flattened if desired by specifying an extra point making the slope of the last segment flat.

Set the independent voltage source V1 as PWL(0 1 50u 6 100u 2 150u 10 200u 8 250u 1). The user can define unlimited pairs of (time, value) sequences using the PWL function. Here, two more data point pairs are added as shown in Fig. 5.29.

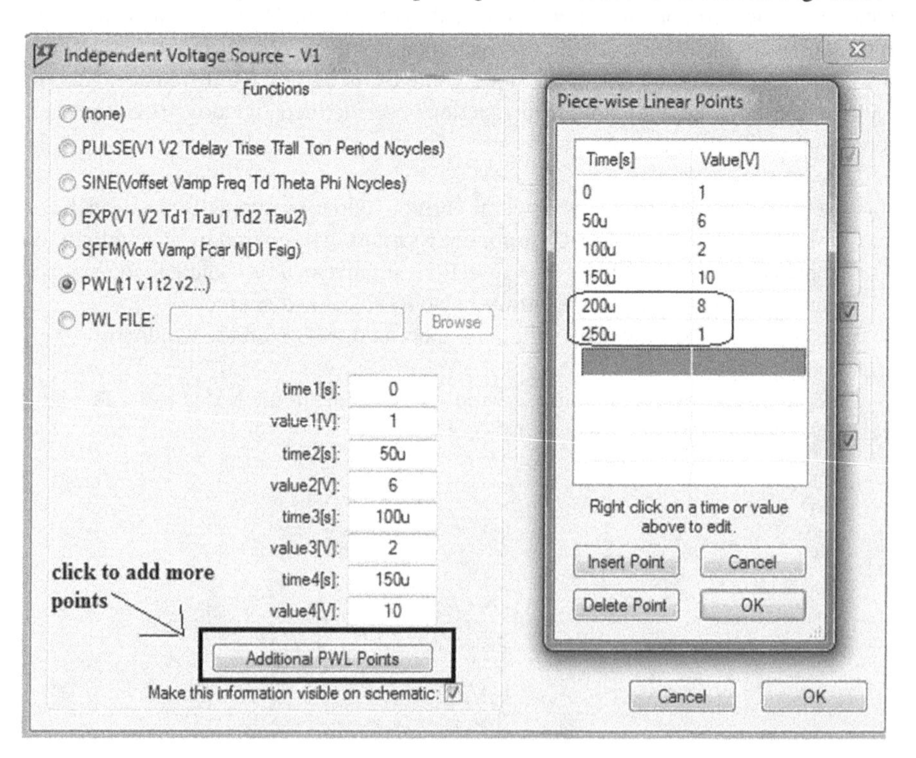

FIGURE 5.29 The PWL Function Setting for the Voltage Source Component.

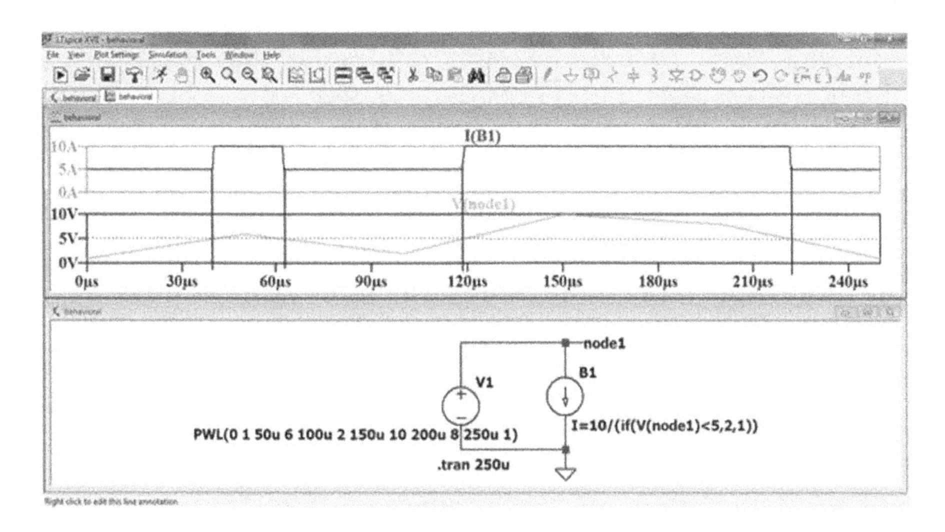

FIGURE 5.30 The Simulated Output.

Also, by selecting the PWL FILE function from the Component menu window, a large number of points can be easily imported into the simulator to generate more sophisticated waveforms.

v. Run a transient analysis for 250 microseconds to obtain the simulation output as shown in Fig. 5.30:

The circuit schematic and output:

An output current through the B1 is 10/2 = 5 A if a voltage at the node1 is less than 5 V, otherwise 10/1 = 10 A.

5.7 VCVS AS OP-AMPS FOR TRANSIENT SIMULATIONS

In a linear region, an OP-Amp output voltage changes linearly with a differential input voltage, and therefore voltage-controlled voltage source can be used to realize an OP-Amp. But to operate in a linear region, an OP-Amp differential input has to be very small (few microvolts) as its open-loop gain is generally in the order of 10^5. A linear region operation of an OP-Amp is achieved by using negative feedback.

6 AC Analysis

6.1 INTRODUCTION TO AC SWEEP SIMULATIONS

An AC analysis in SPICE computes an AC output variable (voltage differences between the specified nodes, a voltage at the node referring/about the common ground, a current flowing out of the independent voltage source, the elements branch currents, impedance (Z), admittance (Y), hybrid (H) and scattering (S) parameters, input/output impedances, and admittances as a function of frequency. The simulator first computes a dc operating point of the circuit and determines the linearized, small-signal models for non-linear devices in the circuit. The resultant linear circuit is then analyzed over a user-specified range of frequencies. The desired output of an AC small-signal analysis is usually a transfer function (voltage gain, trans-impedance, etc.). If the drafted circuit has only one AC input, it is suitable to set the input signal peak amplitude (maximum voltage) to unity and phase to be zero, so that an output variable (voltage) represents the same value as a transfer function of the output variable to the input variable.

An AC simulation analysis sweeps through different frequencies at a constant amplitude to study how an output voltage/current throughout a circuit responds to different driving frequencies. An AC analysis includes representing quantities like impedance or gain-phase as a function of frequency. It requires including a sinusoidal source in the circuit schematic.

The AC analysis is used to compute a gain magnitude or amplitude of the output signal (in dB) and phase difference between the input and output signals as the input frequency changes for circuits like filters, amplifiers, etc., and logarithmic (decade) sweep is preferred mostly. AC analysis is also useful for finding a transfer function, bandwidth, and stability of amplifiers. It allows plotting Bode Plot for circuits. AC analysis uses a small-signal linear model for all non-linear devices (diodes, transistors, etc.) and does not provide meaningful results if these devices operate in a non-linear mode in a circuit.

AC analysis does complex phasor analysis and can also be used to determine complex node voltages and device currents as a function of frequency. All non-linear elements are replaced by linear models to obtain meaningful results. For the analysis, the user needs to use a special setup available for the independent sources. A simulation using an AC analysis requires changing the independent voltage source to an AC signal having fixed amplitude and phase, but varying frequency.

While a DC sweep is simulated by sweeping through different values of a DC voltage/current source, an AC analysis is simulated by sweeping through different frequencies at a constant amplitude.

Thus, an AC analysis plots the magnitude and/or phase response of a circuit versus frequency for different inputs which include:

i. **Type of Sweep**
In the AC analysis menu or fields, the user gets the three analysis options to describe an x-axis scaling and sweep a frequency on a linear, octave, or decade scale. For example, if the decade sweep option is selected, then a sample of x-axis contains 100 Hz, 1 kHz, 100 kHz, 10 MHz, etc. Therefore, to observe how a circuit reacts over a very large range of frequencies choose the decade option.

ii. **Number of Points, Start Frequency, and Stop Frequency**
The user needs to specify how many points the software should use for calculating frequencies and what should be the start and end frequency values.

6.2 AC ANALYSIS OF SERIES RLC CIRCUIT

Create the series RLC circuit schematic as shown in Fig. 6.1 to determine voltages and currents in the circuit elements.

i. Connect the resistor R1, inductor L1, and capacitor C1 model in series across the voltage source component V1. Assign the net name Vin to an input node. Assign the net name Va to the node connecting the R1 and L1 and to the node connecting the L1 and C1 together.

- **Assign Values to the Components:** Set the R1 value to 30Ω, the inductor L1 value to 20 mH, and the capacitor C1 value to 2 μF.

(**Note:** The user should assign values to all the parameter fields in the component attributes editor window, whereas a few information fields (e.g., Series Resistance) are optional and can be left empty. The object's editor window can be opened by moving a cursor over the body of the object, and pressing a right mouse button when a pointing finger appears.)

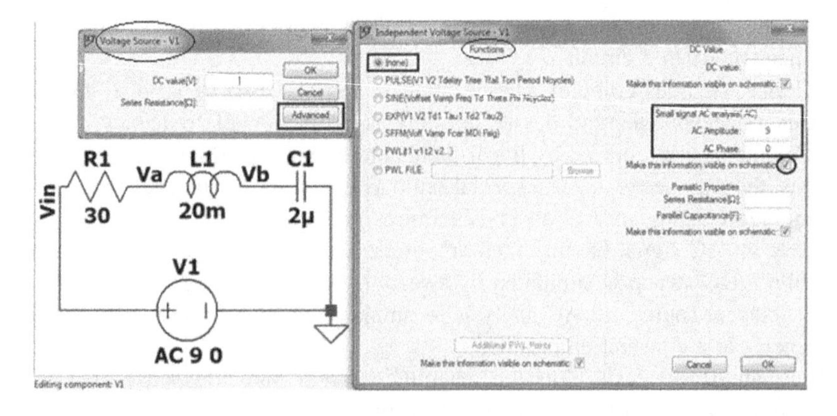

FIGURE 6.1 The Series RLC Circuit Schematic.

ii. For the settings of the voltage source component V1, right-click on the component symbol (or body of the component) to open the value editor window referred to as **Voltage Source – V1**. Now, click on the **Advanced** button (for performing an AC analysis) to open the component menu dialog box and do the entries as shown in Fig. 6.1.

iii. For an AC (frequency response) analysis, select the **(none)** radio button tab from a list of functions on the left-side under the **Functions** settings, then go to the **Small signal AC analysis(.AC)** section on the right-side and enter 9 in a text box next to the **AC Amplitude** field and enter 0 in the **AC Phase** text box. Thus, the AC magnitude and AC Phase values selected for an AC analysis are 9 V and 0°, respectively. It is to be noted that a simulation frequency is not entered at this point and it should be done later. Alternating current and voltage sources are impulse functions used for an AC analysis.

(**Note:** In LTspice XVII, a letter **u** works for a **microfarad** and **m** represents **a millihenry**.)

Remember: When any trouble arises while drafting the circuit, see Section 1.3.5 on Additional features which include commands for information on how to delete, move, mirror, and drag the components, zoom the schematic to a full extent.

IMPORTANT: During an AC analysis, any of the sources can be selected without affecting the results. In an AC analysis, only the AC Amplitude and AC Phase value parameters create any effect on the simulation. All other values are ignored during an AC analysis.

6.2.1 Impedance Computation

- **Using the .meas Directive Statement**

iv. Before running a frequency-domain simulation, an inductive/capacitive reactance or impedance of the circuit at a specified value of $2\pi f$ or w (rad/sec) can be computed according to their formulas using the .meas statement. Place the .meas directive on the circuit schematic using the SPICE Directive schematic tool. Right-click to open the .meas Statement Editor for editing the parameters.

As the formula for capacitive reactance is given by $X_C = \frac{1}{wC}$, the capacitive reactance X_C of the C1 can be computed for the specified value of w, say 1000 rad/sec by using the .meas statement in the parameter format as shown in Fig. 6.2. The **Applicable Analysis** field is chosen to be AC (note that here the (any) option can also be selected), the **Result Name** field is named Xc, the **Genre** field is selected to be PARAM and in the **Measured Quantity** field do the entry $1/(1000 * 2 * 10^{**}(-6))$, where $w = 1000$ rad/sec and $C = 2 \times 10^{-6}\ F$. After completing the entries for the statement, click the OK. Ensure that the complete .meas directive statement is placed on the schematic.

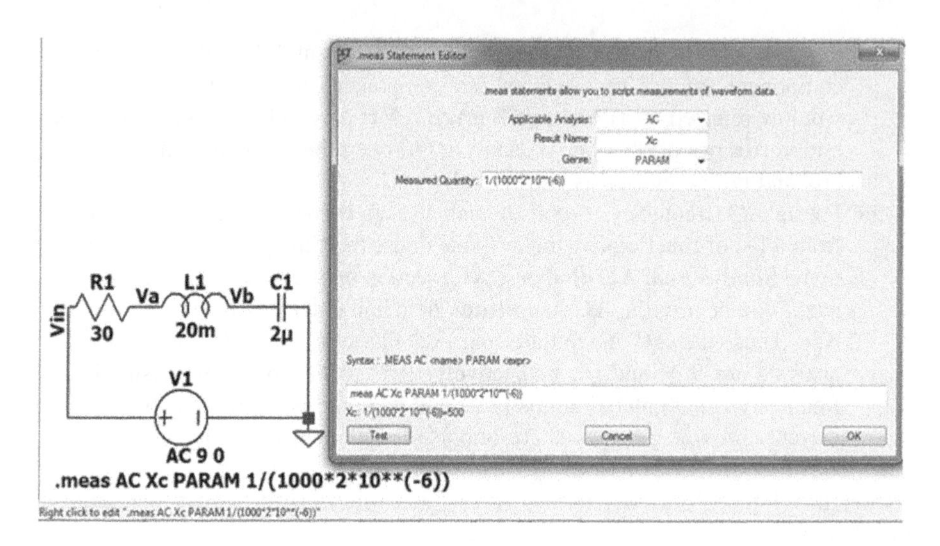

FIGURE 6.2 Measuring the Capacitive Reactance.

Right-click the complete .meas and click the Test button to see the result. The **Test** button on the bottom left-hand side evaluates an expression before the circuit is simulated. The capacitive reactance X_C can be read as 500Ω (see Fig. 6.2).

6.2.2 Total Impedance with the Built-In Functions

LTspice XVII contains many built-in mathematical functions for doing mathematical operations (such as addition, subtraction, multiplication, and division, trigonometry, log, power, square root, integration, and differentiation). The syntax for using the built-in functions takes the form:

built-in_function_name(arg1, arg2)

where the built-in_function_name is the valid command followed by either a single argument enclosed in parentheses like tan(45) or a double argument like pow(x,2).

The syntax for using the .meas for computing the expression using the built-in functions takes the form:

.meas Result_Name PARAM expression_using_built-in_function

The command **.meas tangent PARAM tan(45)** outputs 1, whereas the command pow(x,2) outputs a square of the already defined parameter x.

v. As an impedance of a series RLC circuit is given by a formula $Z = \sqrt{R^2 + (X_L - X_C)^2}$, the total impedance can therefore be computed by using the parameter format as shown in Fig. 6.3:
vi. In the Measured Quantity field, do the entry **sqrt(pow(30,2)+pow((1000*0.02–1/(1000*2*10 **(–6))),2))**, then click the OK and place the complete directive on the schematic.

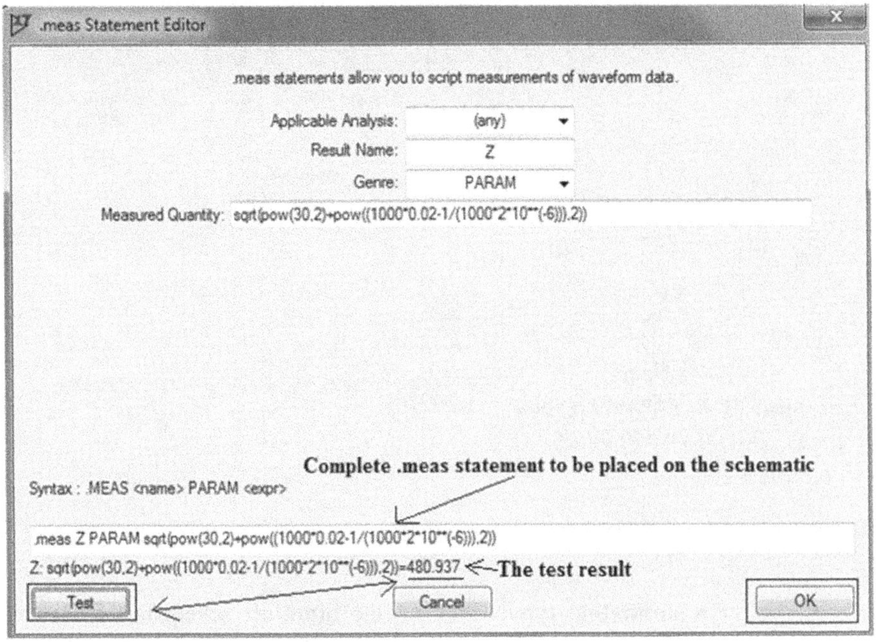

.meas Statement Editor

.meas statements allow you to script measurements of waveform data.

Applicable Analysis: (any)

Result Name: Z

Genre: PARAM

Measured Quantity: sqrt(pow(30,2)+pow((1000*0.02-1/(1000*2*10**(-6))),2))

Complete .meas statement to be placed on the schematic

Syntax : .MEAS <name> PARAM <expr>

.meas Z PARAM sqrt(pow(30,2)+pow((1000*0.02-1/(1000*2*10**(-6))),2))

Z: sqrt(pow(30,2)+pow((1000*0.02-1/(1000*2*10**(-6))),2))=480.937 ← The test result

Test Cancel OK

FIGURE 6.3 Measuring the Total Impedance Using the .meas.

After placing the complete .meas statement on the schematic, right-click on it (Right click to edit ".meas Z PARAM sqrt(pow(30,2)+pow((1000*0.02-1/(1000*2*10**(-6))),2))"), and then click the **Test** button.

The total impedance of the circuit comes out to be 480.937 ohms.

Manually,

$$Z = \sqrt{R^2 + (X_L - X_C)^2} \tag{6.1}$$

Therefore,

$$Z = \sqrt{(30)^2 + (20 - 500)^2} = \sqrt{900 + 230400} = \sqrt{231300} = 480.936586256 \ \Omega$$

Both the theoretical and simulation results agree.

6.2.3 Setting AC Analysis Parameters

vii. Simulate an AC analysis after completing the circuit drawing and assigning parameter values to all the components/devices along with setting up the voltage source V1. It is optional to delete the .meas command using the Cut tool.

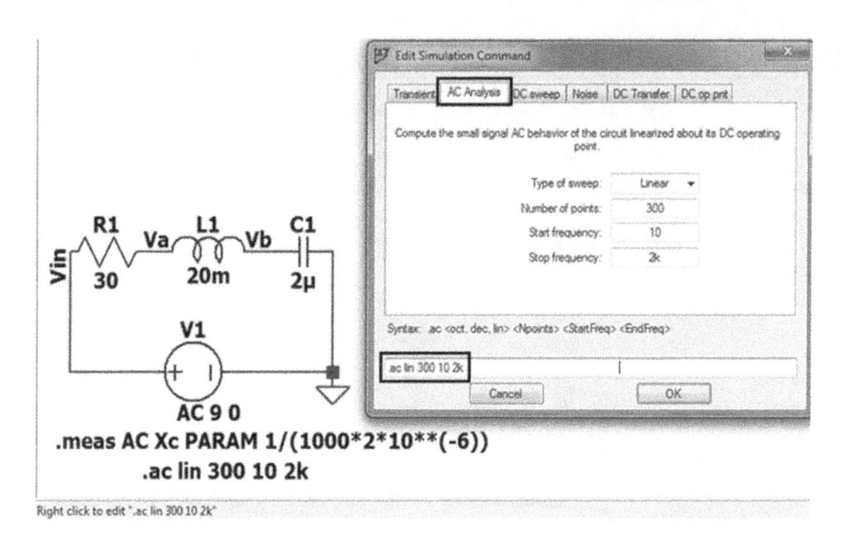

FIGURE 6.4 Setting an AC Analysis Parameters.

- **Select a simulation type:** Click on the **Simulate** schematic menu and choose the **Edit Simulation Cmd** sub-menu to open the **Edit Simulation Command** dialog window. Select the **AC Analysis** tab, then choose a sweep type to be **Linear** from the pull-down menu next to the **Type of sweep** field and set the frequency parameters to perform an AC Analysis on frequencies starting from 10 Hz (**start frequency**) and going up to 2 kHz (**stop frequency**) as shown in Fig. 6.4. The analysis is performed here at a total of 300 (**number of points**) different frequency points.

Click the OK and place the .ac statement on the schematic to start a simulation.

It is to be noted that by default, an AC Analysis calculates the circuit's response versus frequency in Hertz.

To plot a frequency response that is computed between 10 rad/s and 2 krad/s, change the simulation command to:

```
.ac lin 300{10/(2*pi)}{2k/(2*pi)}
```

The simulated result still shows a magnitude versus frequency plot displaying frequency in Hz on a horizontal axis.

To display an angular frequency in rad/s on a horizontal axis, first change the simulation command as shown here:

```
.ac list {w/(2*pi)}
```

then add the following directive to the schematic:

```
.step dec param w 10 2k 50
```

The plot displays an angular frequency parameter on an x-axis.

6.2.4 Running Simulation

viii. Press the **Run** schematic tool button to open the **blank** waveform viewer window (see Fig. 6.5) horizontally above the schematic.

- **How to get information on the circuit (probing and measuring electrical quantities):**
  Probe the Quantities for Plotting. All the node voltages with reference to the common ground (left-click on the node/net/wire), differential voltages or voltages across devices (click + drag between the nets), device currents (left-click on the device/component), and gains (keep AC source = 1V) can be probed and the complex magnitude (in dB units by default) along with the phase response is plotted in the waveform viewer as a function of frequency.

Thus, clicking on the circuit in/at different locations provides information about a voltage, current, and power at/through a specific node/component. Also, clicking on different locations adds more plots to the same plot pane for easy analysis and comparison.

In the plot pane, the user can move a cursor over either a vertical axis or a horizontal axis and left-click a mouse when a cursor changes into a ruler shape. The action pops-up the dialog window where the user can change several options corresponding to the axis. The dialog window for a left-vertical axis allows the user to

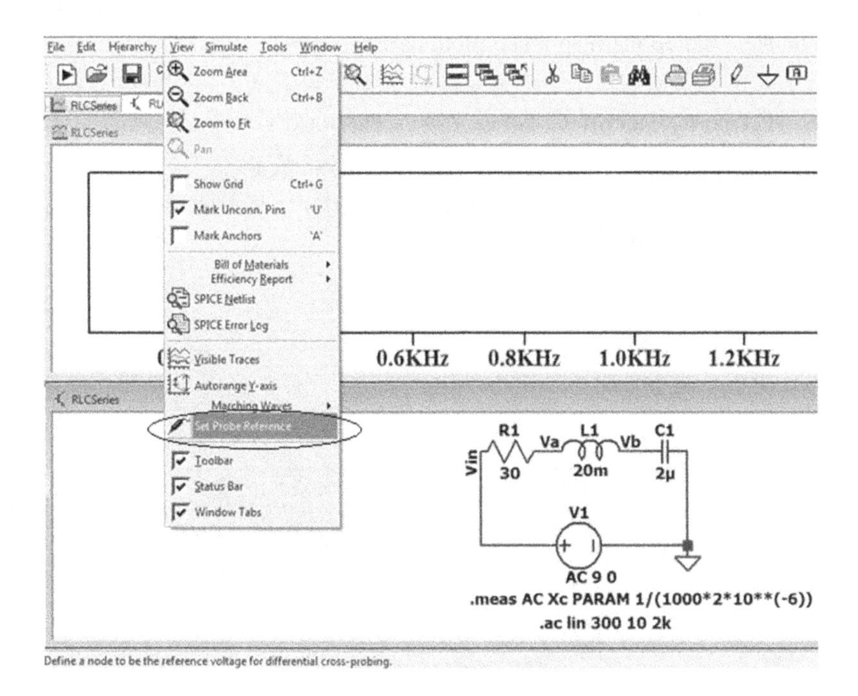

FIGURE 6.5 The Set Probe Reference Sub-Menu under the View Schematic Menu.

choose between Bode (magnitude and phase versus frequency), Nyquist (imaginary component versus real component), or Cartesian (real and imaginary components versus frequency). Further, in Bode (the default) mode, a left-vertical axis can be changed to use a linear, logarithmic, or decibel (the default) scale. A right-vertical axis can be changed to plot a group delay instead of a phase. A frequency (horizontal) axis can be changed between a linear and logarithmic (the default) scale.

6.2.5 Probing using Set Probe Reference Sub-Menu

- **Plotting a voltage across the inductor L1:**

ix. To measure a voltage between two points on the drafted circuit schematic, set a new reference point at one node, say Vb and then click onto another node, say Va. To plot a voltage across the L1, that is, between the Va and Vb, change the reference point from the common ground node to the Vb by selecting the **View** schematic menu and activate the **Set Probe Reference** option by clicking on it as shown in Fig. 6.5.

The transparent voltage probe appears. Bring it at the node Vb and left-click so that the black probe appears denoting that a new reference point is now Vb. Now, release a mouse button and move a probe cursor at the Va. Left-click at the Vin when the red voltage probe appears to plot the V(Va, Vb) as shown in Fig. 6.6.

Thus, a magnitude and phase plot of a voltage across the L1 in the series RLC circuit is plotted as a function of frequency. Right-click on a left-vertical axis (dB by default) to set up the magnitude plotting parameters.

6.2.6 Plotting Circuit Current versus Frequency

x. Delete the previously plotted trace by right-clicking on the trace label V(Va,Vb) to open the **Expression Editor** window and choosing the **Delete this Trace** button.

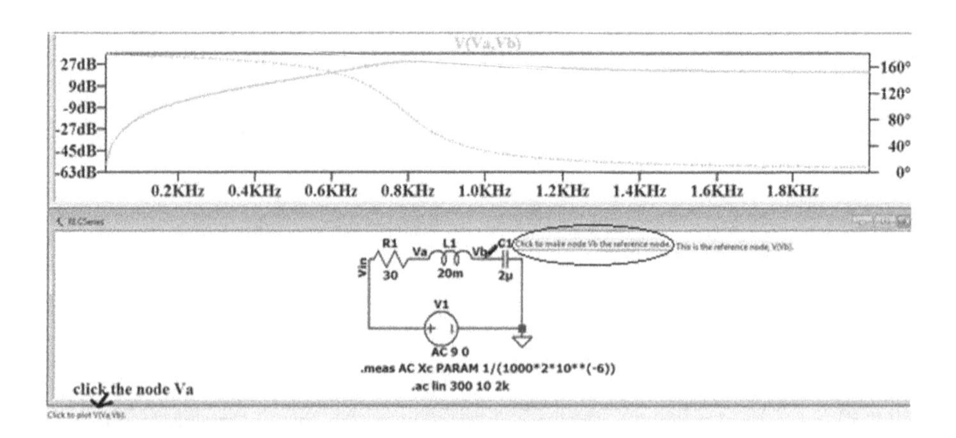

FIGURE 6.6 Differential Voltage Plotting with the Set Probe Reference Sub-Menu.

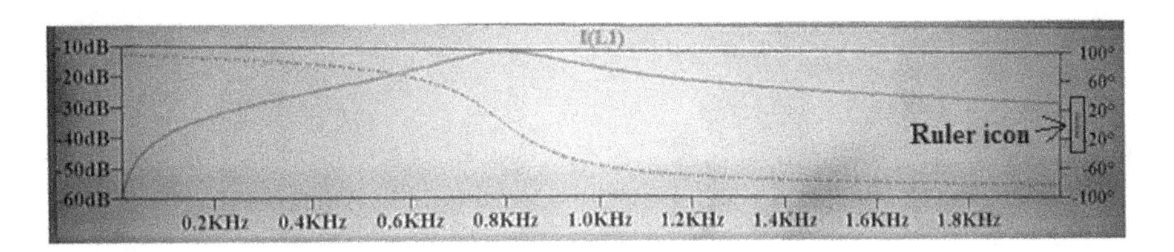

FIGURE 6.7 Controlling the Right-Axis Parameters.

- **Plotting the Inductor Current:**

xi. Now, move a cursor on the component body, say inductor, and left-click to plot the I(L1). After probing using the current sensor, a magnitude and phase plot of the current flowing in the series RLC circuit is plotted as a function of frequency. Hover a cursor along a right-vertical axis outside of the plot until a ruler icon appears as shown in Fig. 6.7, and then right-click for controlling the axis parameters.

- **Delete the Phase Plot (in the default unit degrees)**

xii. It is to be noted that by default both phase and magnitude responses get displayed on the waveform viewer window. A phase plot can be easily deleted if not required by right-clicking on the phase **axis** (a right vertical axis)at the right side of the plot pane and clicking on the **Don't plot phase** button in the Right Vertical Axis dialog window that pops up as shown in Fig. 6.8.

Now, press the OK to remove the phase plot and retain only the magnitude trace plotted in the default unit of decibel.

- **Editing a Magnitude Axis**

xiii. When the waveform viewer window is visible, move a cursor over a left-vertical axis (i.e., along a magnitude axis on the left side of the plot pane) outside of a plot and the cursor transforms into an image of a ruler. Now,

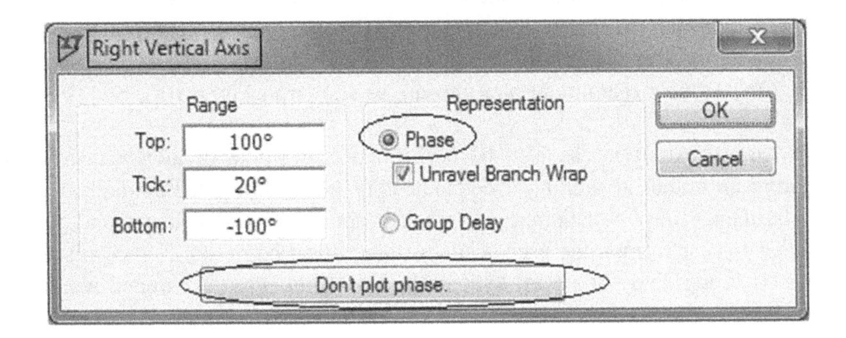

FIGURE 6.8 The Dialog Window for the Phase Axis (Right-Vertical Axis).

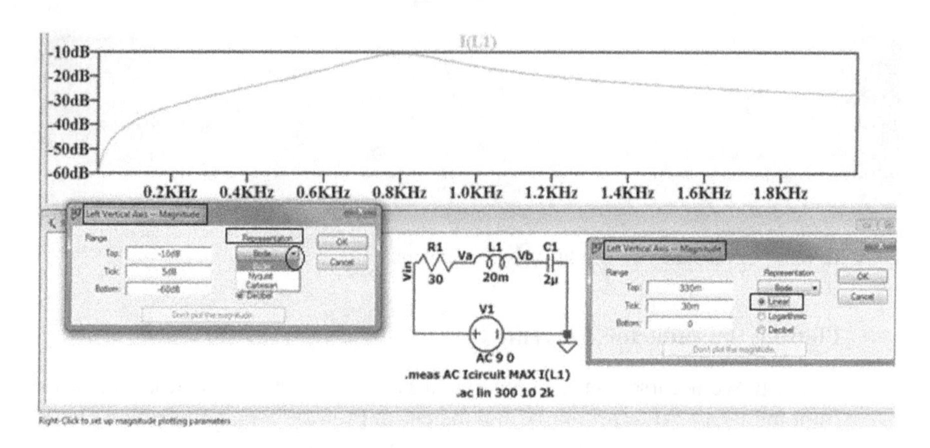

FIGURE 6.9 The Magnitude Dialog Window for Left-Vertical Axis.

right-click a mouse to open the **Left Vertical Axis -- Magnitude** dialog window to change several options corresponding to the axis. The dialog window for a left-vertical axis (see Fig. 6.9) allows the user to select between Bode, Nyquist or Cartesian plot.

- **The circuit current on a linear scale:**
  In Bode (the default) mode, we can further change the left vertical axis to use a linear, logarithmic, or decibel (the default) scale. Similarly, a right-vertical axis can also be edited to plot a group delay instead of plotting a phase. A frequency (horizontal) axis can also be changed between a linear and logarithmic (default) scale in the same way.

Thus, an axis (left-vertical, right-vertical, or horizontal) parameter values (Top/Left, Tick/tick and Bottom/Right) of a plot can be changed by hovering a cursor along the axis outside of the plot until a ruler icon appears, and then right-click. Similarly, a family of plots (e.g., an amplitude or phase for a Bode plot) displayed can be deleted by right-clicking on the plot label/name.

xiv. In a **Bode** (the default) mode for representing a plotted trace, change a left-vertical axis to use a linear scale by checking the **Linear** button option under the **Representation** field pull-down menu (see Fig. 6.9). Clicking the OK, a linear magnitude plot appears as shown in Fig. 6.10.

(**Note:** An AC Analysis can also be performed at a single frequency instead of computing an output at multiple frequencies. To perform a complex analysis at a single frequency in phasor domain circuits for obtaining node voltages and device currents (in polar forms), the user needs to select the **List** option in the pull-down menu next to the **Type of sweep** field in the Edit Simulation Command window.)

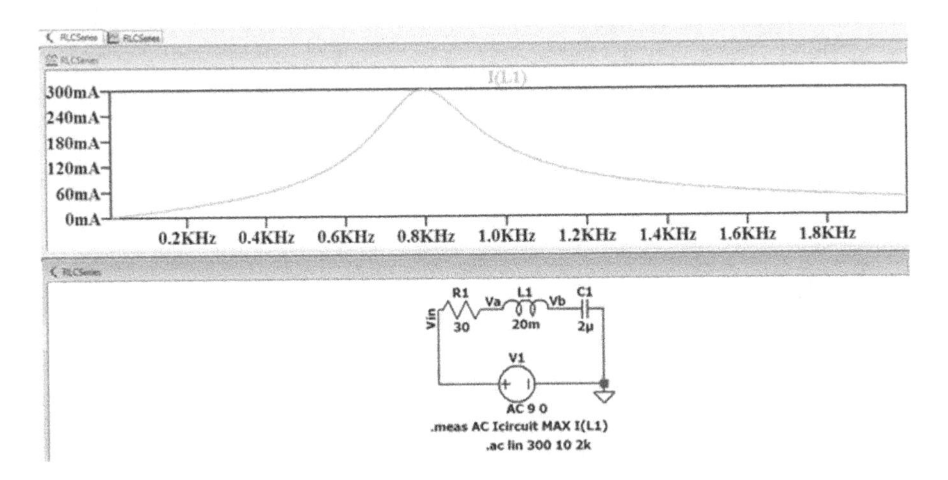

FIGURE 6.10 The Circuit Current Linear Plot.

Remember: The user can change the axis parameters of a traced plot by hovering a mouse cursor along an axis (vertical or horizontal) outside of the plot until a ruler icon appears and then doing a right-click. This action allows for changing axis values. In a case of a vertical axis, a family of plots such as an amplitude or phase response for the traced Bode plot can be deleted. Also, the deleted phase response plot can be brought back by right-clicking on a right-vertical axis of the plot.

6.2.6.1 Graphical Measurements

- **The maximum circuit current using the .meas statement:**
  After probing the circuit current, a maximum current through the circuit can be measured for the plotted waveform by using the .meas directive statement. It is to be noted that the current magnitude in an AC series circuit flowing through all the components is the same but a phase value is different for the reactive components.

xv. To evaluate the maximum circuit current, first, place an empty .meas statement on the schematic. Then, right-click on the .meas and do the entries to complete the statement as follows:

Enter a unique name in the **Result Name** field text box, for example, Icircuit.

Click on the drop-down menu in the **Genre** field having many options including AVG, PP, MAX, MIN, and RMS. Select the **MAX** option.

In the **Measured Quantity** field text box, enter I followed by the component name, that is, I(L1).

For precision, the completed .meas statement can be seen in the last field of the Statement Editor window (see Fig. 6.11).

Click the OK so that the statement gets placed on the schematic. Right-click on the full .meas statement already placed on the schematic to re-open the Statement

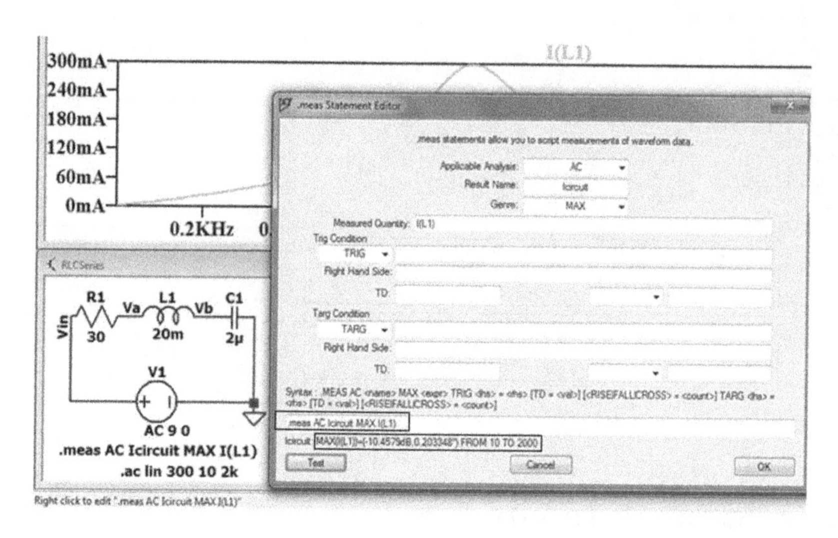

FIGURE 6.11 The Maximum Circuit Current Using the .meas.

Editor window. Left-click on the **Test** button to read the measured result as shown in Fig. 6.11.

Thus, it can be seen that a maximum value of a current in the circuit is equal to −10.4579 dB (i.e., 0.299989 µA or 299.989 mA).

If calculated manually,

$$I_{max} = \frac{V\,(peak \;\; value \;\; of \;\; supply \;\; voltage)}{R} \tag{6.2}$$

$$\therefore \;\; I_{max} = \frac{9}{30} = 0.3 \;\; A \;\; or \;\; 300 \;\; mA = maximum \;\; circuit \;\; current$$

This shows that both the experimental and theoretical results match each other.

- **The Circuit Current at a Specified Frequency**

xvi. Similarly, the .meas directive can also be used for measuring the magnitude of a current (or any electrical quantity) when a certain frequency, say 795.6 Hz is reached.

The circuit current at a frequency of 795.6 is measured to be equal to −10.4588 dB (i.e., 0.299958 µA or 299.958 mA) with the following entries in the .meas directive statement editor window (see Fig. 6.12):

Applicable Analysis: AC, **Result Name**: I, **Genre**: FIND, **Measured Quantity**: I(L1), under the **Point** field, select **AT** from the pull-down menu and type **795.6** in a text box next to the pull-down menu.

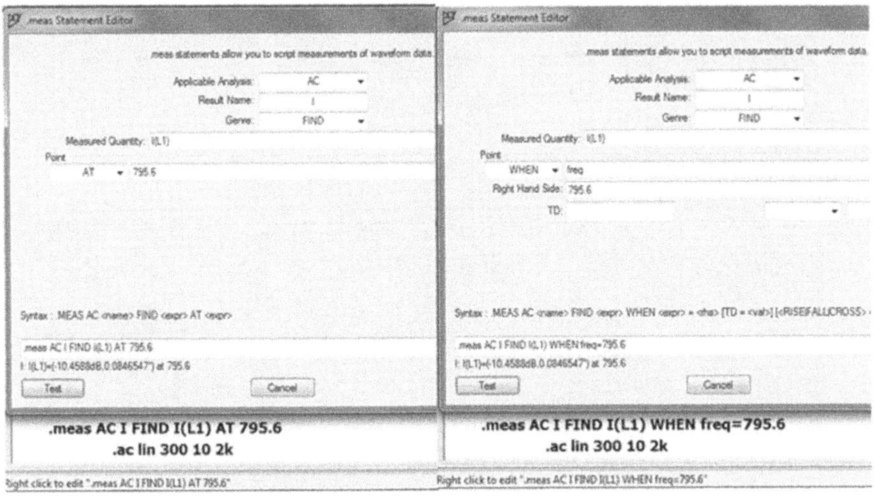

FIGURE 6.12 The Circuit Current at a Specified Frequency using the .meas.

or

Applicable Analysis: AC, **Result Name**: I, **Genre**: FIND, **Measured Quantity**: I(L1), under the **Point** field, select **WHEN** from the pull-down menu and type **freq** in a text box next to the pull-down menu, **Right Hand Side**: 795.6.

- **Frequency Measurement Using the LTspice Numbered Cursors**
  By employing the numbered cursor, the user can measure a frequency (on an x-axis) and an amplitude (on a y-axis) of any particular point on the graph.

xvii. After plotting the current trace, left-click on the trace-title label at the top of the waveform viewer window to set a single cursor cross-bar and popping up the new floating cursor window with information of the cursor position/location as shown in Fig. 6.13. Move a mouse pointer over the new cross-bar till a number 1 appears which represents the Cursor 1. Now, left-clicking the number to move the cursor along with the plotted wave (**the Cursor 1 follows the waveform**) and position it as desired. To remove the cursor just close the little floating information window. To attach the two cursors, right-click over the trace label/name.

Frequency Value when the Circuit Current is Maximum: Right-click over the name of the waveform, that is, I(R1), and choosing an option to display (or left-click for attaching directly). Now, move a mouse over the cross-hairs until the yellow 1 appears, and then drag the cross-hairs left (or right) while holding a left mouse button down till the cursor window displays a magnitude value (vertical y-axis value) of 299.989 mA (maximum circuit current) in a text box next to the Mag field. Now, note down a frequency value in a text box next to the Freq field in

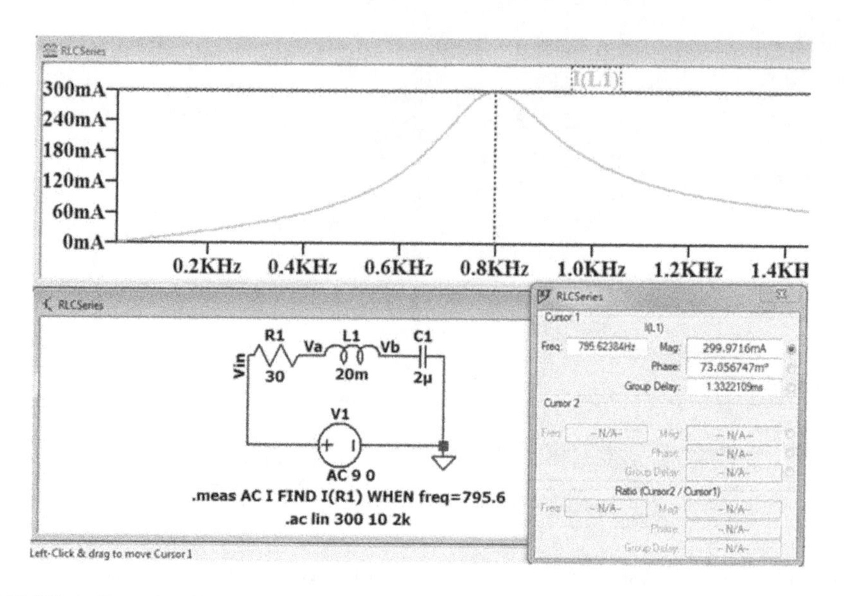

FIGURE 6.13 The Circuit Current at a Specified Frequency using the Cursor 1.

the floating cursor window. A frequency when the circuit current is maximum can be read as 795.6 Hz (after rounding) (see Fig. 6.13).

A frequency (x-axis value) at the magnitude value (y-axis value) of 299.989 mA which corresponds to a maximum current in the circuit can be read as 795.62384 Hz (which corresponds to a resonant frequency of the circuit) in the Frequency box of the little cursor window.

Thus, as a maximum current of 299.9716 mA flows in the series RLC circuit (or through the series elements) at a frequency of 795.62384 hertz, it can be said that resonance in the circuit is obtained at 795.62384 Hz.

Mathematically, the resonant frequency f_r is given by:

$$f_r = \frac{1}{2\pi\sqrt{LC}} \tag{6.3}$$

$$\therefore f_r = \frac{1}{2\pi\sqrt{20\times 10^{-3}\times 210^{-6}}} = 796 \text{ Hz}$$

Thus, the experimental result agrees with the mathematical value.

- **Computing the –3 dB Bandwidth:**

xviii. **Enabling both the Cursor 1 and Cursor 2 on the same wave/trace:** Permit the two cursors on a trace I(L1) by selecting the option **1st & 2nd** from the drop-down menu next to the Attached Cursor field box to reading a difference between their positions in a vertical dimension (to measure a difference between the amplitude levels, say –3 dB) as well as a difference between their positions in a horizontal dimension (to measure a difference between the frequency values when the amplitude falls by a specified value, say –3 dB or in linear terms, the amplitude becomes 0.707 times the maximum value). As the circuit maximum current is 300 mA, so the circuit bandwidth is defined as the difference between the upper –3 dB cut-off frequency point and lower –3 dB cut-off frequency point where the current amplitude becomes 0.707 times 300 mA, that is, 212 mA. Thus, the –3 dB point (also called a half-power point) is defined as the frequency at which the system's output/gain magnitude is reduced to 0.707 of its maximum value.

Position both the cursors so that their respective text boxes next to the Mag field read 212 mA. Now, note down the difference between the positions of Cursor 1 and Cursor 2 in the Freq text box under the Ratio (Cursor2 /Cursor1) field (see Fig. 6.14) which corresponds to a –3 dB bandwidth of the circuit. Thus, it can be stated that a –3 dB bandwidth of the circuit is 239 Hz.

Mathematically, a –3 dB bandwidth ($BW_{-3\,\text{dB}}$) is computed as:

$$BW_{-3\,\text{dB}} = \frac{R}{2\pi L} \tag{6.4}$$

$$\therefore BW_{-3\,\text{dB}} = \frac{30}{2 \times 3.14 \times 20 \times 10^{-3}} = 238.85 \text{ Hz}$$

Thus, both the experimental and manual results agree.

(**Note:** By right-clicking on the trace name, the user can attach two cursors (choosing the 1st & 2nd) so that a difference between respective data points x and y values can be read in the little window that pops up. Enabling the two cursors by

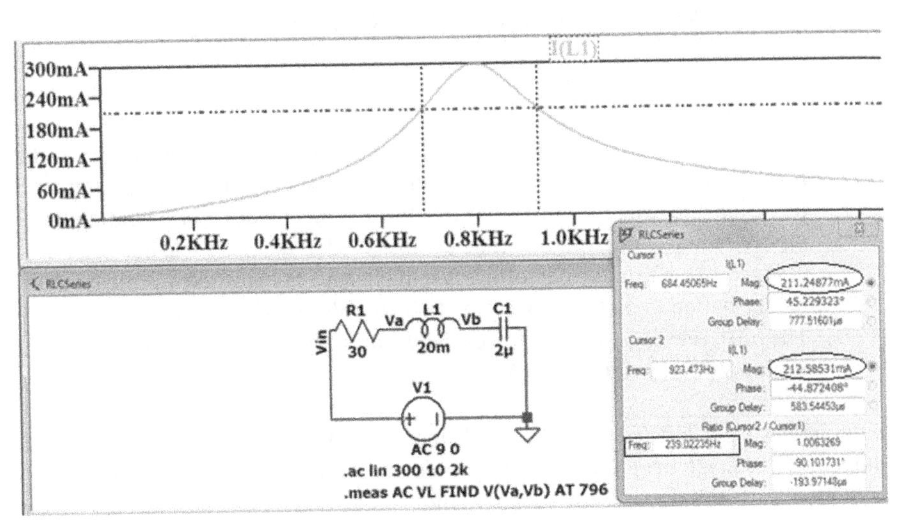

FIGURE 6.14 Computing the Circuit –3 dB Bandwidth.

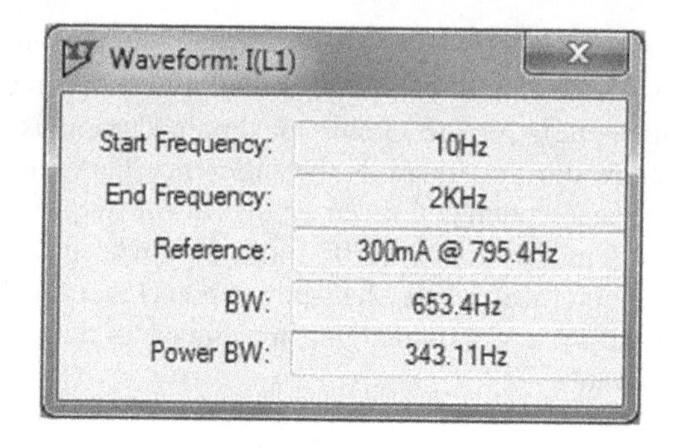

FIGURE 6.15 The Reference Values in Case of a Bode Plot.

selecting the option 1st & 2nd from the drop-down menu next to the Attached Cursor field box helps in reading a difference between their positions in a vertical dimension, for example, to measure peak-to-peak amplitude value as well as a difference between their positions in a horizontal dimension, for example, to measure a time difference Δt or phase difference ($360^0 \times \frac{1}{T} \times \Delta t$) between two waves, the bandwidth of an amplifier, or filter. Also, by right-clicking on the label of another trace, the user can attach the 2nd cursor (Cursor 2) to check the x and y values of the signal in the same plot pane.)

xix. Also, control-click the waveform label/name at the top of the screen to get the reference value and other calculated values as shown in Fig. 6.15.

6.2.7 Differential Voltage Between the two Nodes/Nets versus Frequency

- **Plotting a Voltage across the Inductor Component:**

xx. A voltage across the L1 is equal to the difference of voltages at the nodes Va and Vb. Therefore, for plotting a difference between two voltages at the nodes Va and Vb with the voltage probe cursor, move a cursor at Va (the node of interest) till the red probe cursor appears (see Fig. 6.16), afterward left-click and drag the cursor to the Vb (reference node) while holding a left mouse button. Initially, the probe cursor is red at the node Va, but when it is dragged to the referenced voltage Vb, it changes to black. Now, release a left mouse button to plot the differential voltage Va–Vb (in a dB unit by default). The difference of two voltages is written equivalently as V(Va,Vb) or V(Va)-V(Vb) in LTspice XVII.

After releasing a left mouse button when the black voltage cursor appears, the inductor L1 voltage magnitude (or magnitude of a differential voltage), say Va –Vb is plotted as a solid line, whereas a phase plot is a dashed line. A left-vertical axis is

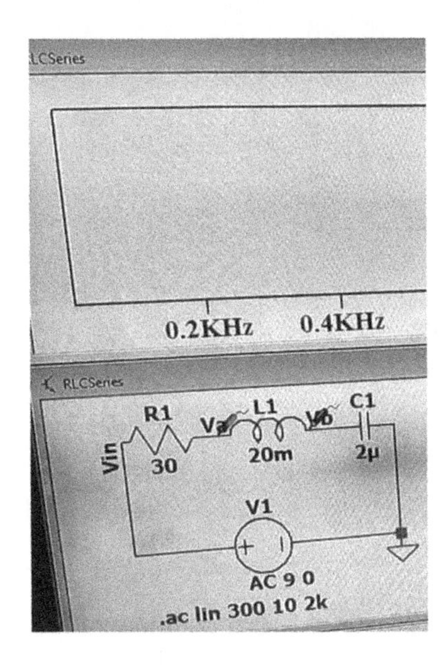

FIGURE 6.16 Differential Voltage Plotting.

the magnitude axis while a right-vertical axis is the phase angle axis. The magnitude and phase plots can also be graphed in separate plot panes.

(**Note:** V(Va,Vb) is complex and by default, the magnitude is plotted in decibels (dB) and the phase angle gets graphed in degrees.)

- **A Voltage across the Inductor at a Specified Frequency**

xxi. Using the .meas directive statement editor window (see Fig. 6.17), a voltage across the L1 can be measured at a frequency of 795.6 Hz. It can be easily seen that at the circuit resonant frequency 795.6 a voltage across the inductor is measured as 29.539 dB or **29.988 V** (very much similar to a voltage value computed manually across the L1 at the circuit resonant frequency).

Mathematically:

- From Eq. (6.3), the resonant frequency $f_r = \frac{1}{2\pi\sqrt{LC}} = \frac{1}{2\pi\sqrt{0.02 \times 2 \times 10^{-6}}} = 796.178$ Hz
- From Eq. (6.2), the peak value of the circuit current at resonance, that is, I_{max} is equal to

$$I_{max} = \frac{V\,(peak\ \ value\ \ of\ \ supply\ \ voltage)}{R} = \frac{9}{30} = 0.3\ A\ \ or\ \ 300\ \text{mA}$$

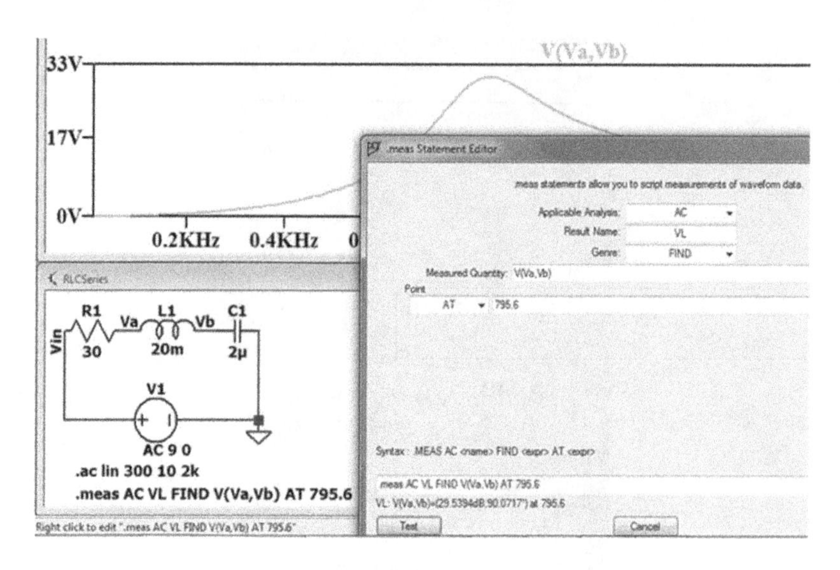

FIGURE 6.17 Measuring Voltage at a Specified Frequency.

- Inductive reactance X_L at resonance is given by:

$$X_L = 2\pi f_r L \tag{6.5}$$

$$\therefore X_L = 2\pi \times 796 \times 0.02 = 100 \ \Omega \ \text{(at resonance)}$$

- A voltage across the inductor at resonance is given by:

$$V_{L\,(max)} = I_{max} X_L \tag{6.6}$$

Therefore, the voltage across the inductor at resonance is equal to $V_{L\,(max)} = I_{max} X_L = (300 \ mA \times 100 \ \Omega) = 30$ V (almost equal to a value 29.988 V computed experimentally).

Thus, it can be stated that the simulation results are almost identical to the mathematically computed electrical values at resonance.

- **Reading Values with the numbered cursor:**

xxii. Right-click the V(Va,Vb), a waveform or trace name, and attach the 1st cursor from the drop-down menu next to the Attached Cursor field. Move the Cursor 1 around with a mouse to read individual frequencies as well as observe magnitude, phase, and group delay at a specific frequency. The already plotted trace can also be deleted by clicking the **Delete this Trace** button.

6.2.7.1 Doing Math for Computing Impedance/Reactance at Resonance

- **Plotting an Inductive Reactance versus Frequency**

xxiii. Enter an expression for plotting an inductive reactance $X_L = \frac{V(Va, Vb)}{I(L1)}$ as follows:

Right-click the waveform name (to do mathematics on the plotted trace) and type in the Expression Editor window as shown in Fig. 6.18 for dividing the V(Va,Vb) by the I(L1) to plot the inductive reactance as a function of frequency (see Fig. 6.19).

- **Measuring the Inductive Reactance at a Resonant Frequency of 795.6 Hz An inductive reactance X_L at the circuit resonant frequency can be observed and verified easily.**

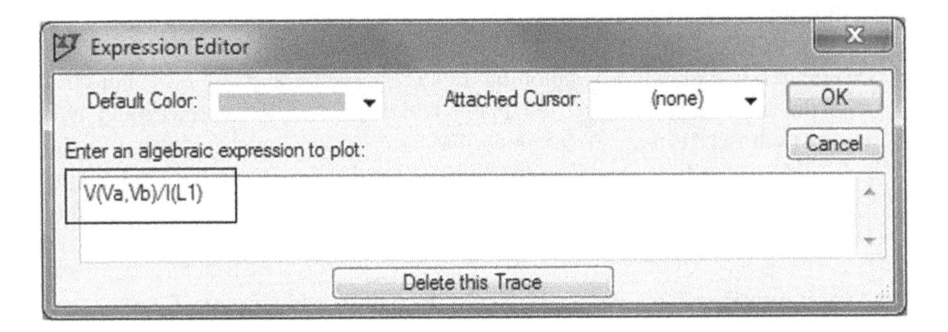

FIGURE 6.18 Entering an Expression in the Expression Editor Window.

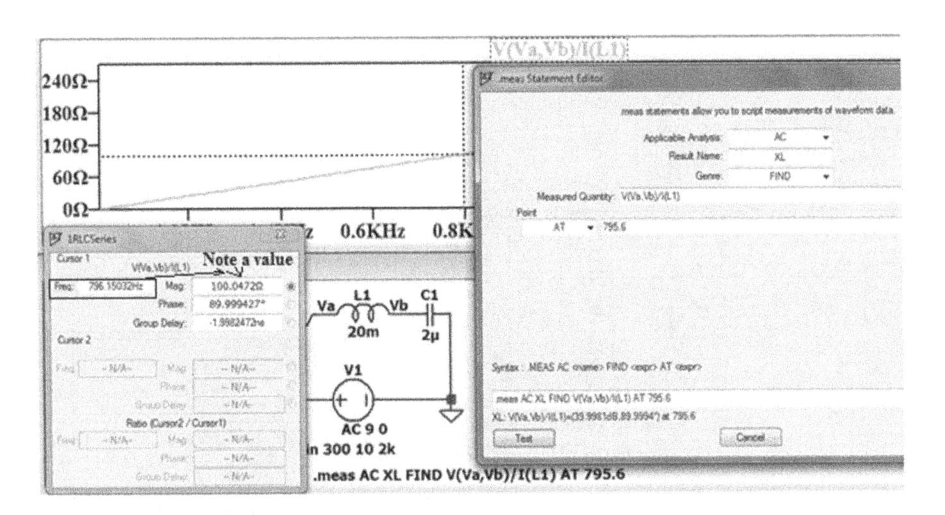

FIGURE 6.19 Computing X_L at Resonance from the Circuit Inductive Reactance Plot.

xxiv. **Use the numbered cursor:** Read the inductive reactance value of the L1 at the resonant frequency 795.62384 Hz using the numbered cursors. Attach Cursor 1 to the wave and drag it about with a mouse so that the readout display (little floating cursor window, see Fig. 6.19) shows a frequency value of 795.62384 Hz in the respective Freq box. At this location, read a magnitude value in the Mag box visible on a screen.

xxv. **Use the .meas statement:** Place the complete .meas directive statement to read the X_L.

Thus, from Fig. 6.19, X_L at resonance can be read as 100.0472Ω or 39.9981 dB (99.978128 ohms).

Manually from Eq. (6.5), the inductive reactance at a resonant frequency of 796 hertz is computed as $X_L = 2\pi \times 796 \times 20 \times 10^{-3} = 100\ \Omega$.

Thus, it can be stated that the simulation result is very much similar to the manually computed value.

- **Plotting the Circuit Impedance:**

xxvi. Enter an expression for plotting a total impedance $Z = \frac{V(Vin)}{I(L1)}$. Right-click the waveform name (to do mathematics on the plotted trace) and type V(Vin)/I(L1) in a text box under the Enter an algebraic expression to plot field in the Expression Editor window for dividing the total input supply voltage by the circuit current to plot a total impedance Z as a function of frequency.

- **Computation of** Z at resonance **using the Numbered Cursor:** Z can be read as 30Ω (see Fig. 6.20)

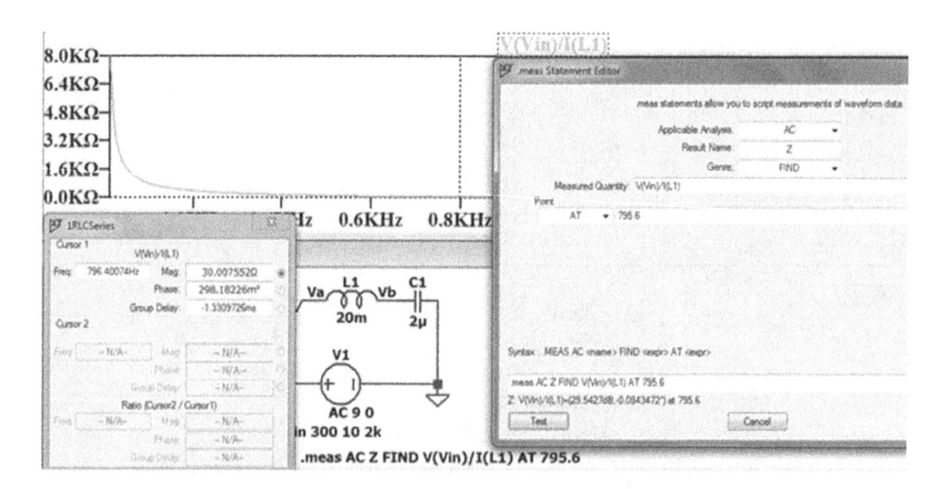

FIGURE 6.20 Plotting the Circuit Impedance along with Computing the Same at Resonance.

- **Measurement of** *Z* at resonance **using the .meas Statement:** *Z* can be read as 29.5427 dB (30Ω)

Mathematically, a total impedance *Z* of the circuit at a manually computed resonant frequency 796 hertz is equal to a resistance value of the resistor in the circuit, that is, $Z = R = 30\ \Omega$.

And, the simulation results show that $Z = V(vin)/I(L1) = 30.007552\Omega$ (or 29.5427 dB = 30.00095Ω) at the circuit resonant frequency 795.62384 Hz which is very much similar to the manually obtained value.

6.2.8 Impedance using Current Source

- **An Alternative Method for Computing an Impedance of the Circuit:**

xxvii. Also, because impedance is simply defined as voltage divided by current (Z = V/I), take I = 1 A (i.e., change the constant current source I1 to an AC version and assign 1 A current to the AC amplitude parameter value) so that the V becomes = to Z. Thus, a voltage can be easily measured/probed in LTspice XVII that essentially equals to an impedance/reactance/resistance by connecting the current source component across the series RLC circuit as shown in Fig. 6.21:

As the current source (AC and DC) symbol has an arrow, always point the arrow in the direction of conventionally flowing current to get accurate results. An interesting little feature under the **markers** menu is the ability to add markers to the circuit to see where the current and voltages have imaginary values in the circuit as well as to observe a phase of the source.

From Eq. (6.5), an inductive reactance at resonant frequency 796 hertz comes out to be equal to 100 Ω. The simulation results obtained attaching the Cursor 1 (see Fig. 6.21) on a trace of the voltage across the inductor L2 using the voltage probe

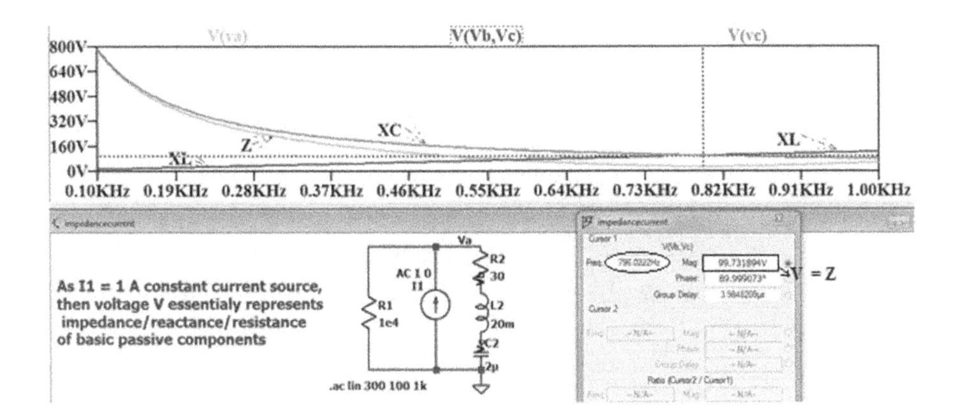

FIGURE 6.21 Impedance using the Current Source Component.

cursor, show that inductive reactance is equal to 99.73189 ohm (as here, V = Z or $V_{L2} = V_b - V_c = X_L$) which is very much similar to the manually computed value.

6.3 BODE PLOT OF TRANSISTOR AMPLIFIER

- **Gain Magnitude Plot of a Common-Emitter Transistor with an Emitter Resistance**

To compute a small-signal gain of an amplifier with coupling capacitors, a DC sweep analysis cannot be used because capacitors are open-circuits at DC. Therefore, an AC analysis is employed to find a small-signal voltage gain within a frequency range. This analysis is performed around an operating point calculated automatically (using the .op analysis without specifying it) and it is the same as a manual small-signal analysis. Since the analysis is the same linearization that is used for hand analysis, there is no saturation for large input voltages.

If the input AC source is set equal to **1**, then the output voltage is numerically equal to the voltage gain.

i. Draw the circuit schematic of a common-emitter npn transistor amplifier with an emitter resistance as shown in Fig. 6.22. Save the circuit diagram with a name npnAC.asc.

(**Note:** Alternatively, open the already saved circuit schematic file npntransient.asc (as the circuit schematic is same) and then save it with a name npnAC to create the schematic file npnAC.asc. Delete the .tran analysis statement already placed on the schematic using the Cut schematic tool. Right-click on the **SINE(0 1 5)** to edit the value of the V1 and type **V** in the **Enter new value for V1** dialog window. Click the OK. Now right-click on the body of the component V1. Under the **Small signal AC analysis(.AC)** setting, set an AC voltage amplitude value = 1 in a text box next to the **AC Amplitude** field, enter a 0 in the **AC phase** box and right-tick

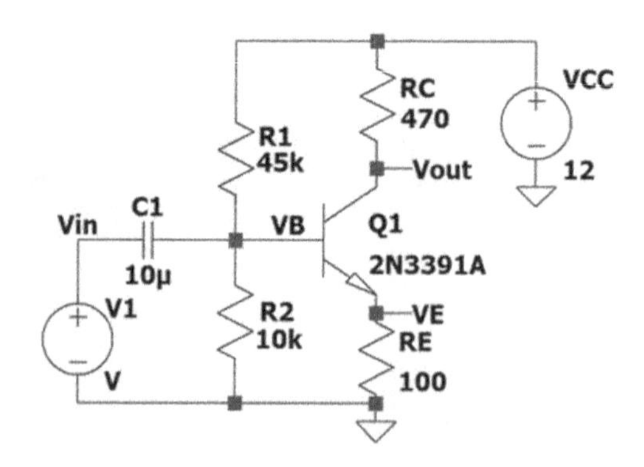

FIGURE 6.22 The Circuit Schematic of a Transistor as an Amplifier.

(or check) the **Make this information visible on schematic** check box. Ensure that the (none) radio button is selected.)

ii. In the earlier-drafted file npnAC.asc, set **the independent voltage source V1 for an AC analysis:** Use the **Advanced** button in the component value editor window to open the component menu window. Select the **(none)** option under the **Functions** setting and set an AC voltage amplitude value = 1 in a text box next to the AC Amplitude field under the **Small signal AC analysis(.AC)** setting. Enter a 0 in the AC phase box.

   To hide the source parameters information, un-tick the **Make this information visible on schematic** check box.

iii. **Set the Parameters for an AC Analysis:** To simulate an AC analysis, set the analysis parameter values as shown in Fig. 6.23. Select the AC Analysis tab in the Edit Simulation Command window. Enter in the respective text boxes:

   Type of Sweep: Decade
   Number of points per decade: 200
   Start Frequency: 0.1
   Stop Frequency: 1000meg

iv. Press the Run schematic tool icon to run an AC analysis simulation. A blank graphical window appears. Probe a voltage at the output node Vout using the voltage probe cursor for plotting a gain (dB) versus frequency (log scale) response output plot as shown in Fig. 6.23.

v. **Delete the Phase Plot:** In the earlier-traced plot, the phase plot is deleted. Information of the circuit's gain (or loss) at each frequency point assists in understanding how properly (or poorly) the circuit can distinguish between signals of different frequencies.

- **The Maximum Circuit Gain (mid-frequency gain) on the Numbered Cursor Readout Display:**

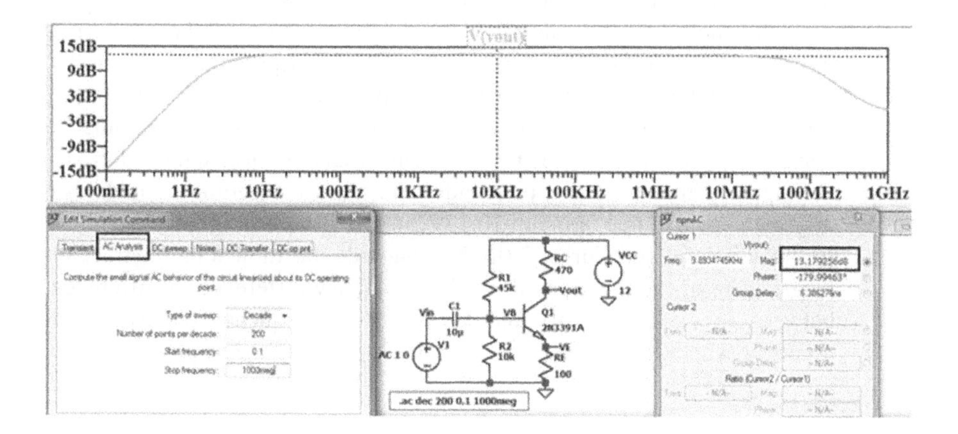

FIGURE 6.23 The Gain Magnitude Plot versus Frequency.

vi. From Fig. 6.23, it can be stated that the voltage gain magnitude of the drafted circuit is 13.179256 dB or 4.560002 V (vertical y-axis position which can be read in the Mag box in the cursor window) at a frequency of 9.8934 kHz (horizontal x-axis position which can be read in the Freq box in the cursor window). Thus, the maximum gain or mid-frequency gain of the drafted transistor amplifier circuit using the Cursor 1 (see Fig. 6.23) is 13.179256 dB. The circuit maximum gain can also be computed using the .meas statement. Place the .meas directive on the schematic as follows: .meas AC maxgain MAX V(vout)

```
.meas AC maxgain MAX V(vout)
```

Leave the Trig Condition and Targ Condition fields unspecified. Press the **Test** button to see the output

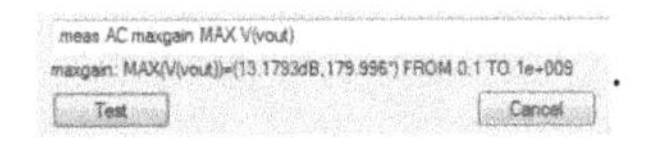

vii. If a trace is plotted by entering the V(vout)/V(Vb) or V(Vout)/V(VB) in the Expression Editor, the voltage gain magnitude comes out to be 13.179276 dB at low frequencies. The circuit gain magnitude agrees with the gain magnitude of the amplifier computed in Section 5.4.1.

6.3.1 Bandwidth Computation

- **The Circuit Bandwidth Measurement:**

viii. Attach both the numbered cursors to a single Bode plot displaying the amplifier gain magnitude against frequency by right-clicking on the trace label V(vout) and selecting the **1st & 2nd** option from the **Attached Cursor** drop-down box to set two cursor cross-bars and popping up the new floating cursor window with information of both the cursors position/location. Move a mouse pointer over the new cross-bar till a number 1 appears which represents the Cursor 1. Drag the Cursor 1 to the left so that the Mag box under the Cursor 1 setting should display a 10.179256 dB gain value which is 3 dB lower than the maximum value 13.179256 dB. Similarly, move a mouse pointer over another cross-bar till a number 2 appears which represents the Cursor 2, and afterward drag the Cursor 2 to the right so that the Mag box under the Cursor 2 setting should also display a gain value around 10.179256 dB. Thus, drag the Cursor 1 (to the left-side) and Cursor 2 (to the right-side) positioned on the trace to locate them at the positions to read a frequency difference (a difference between the upper cut-off and lower cut-off frequency) between the points at which the plotted gain is **3 dB** lower than the maximum value

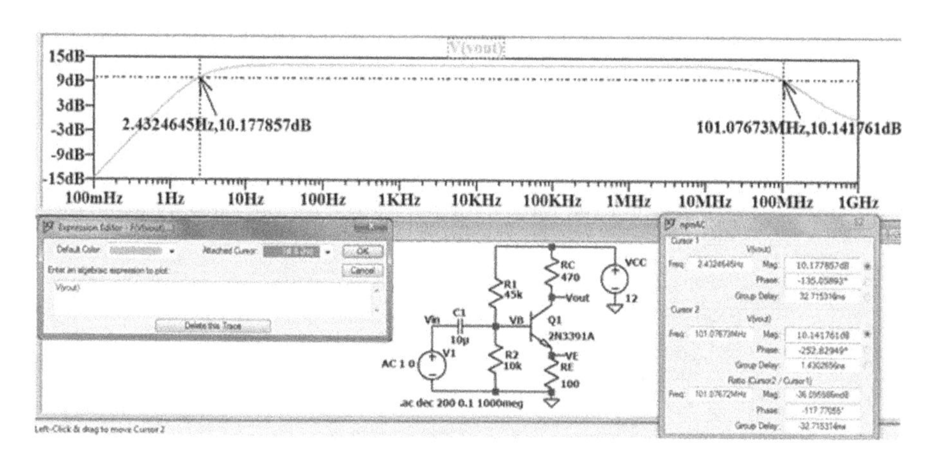

FIGURE 6.24 Reading the Circuit Bandwidth with the Two Cursors.

13.179256 dB, that is, the Mag box should display a 10.179256 dB gain value as shown in Fig. 6.24:

Also, label the cursors positions (go to the Plot Settings -> Notes & Annotations plot/trace/waveform viewer menu/drop-down sub-menu). Afterward, read the display value in a box next to the Freq field under the **Ratio (Cursor2 / Cursor1)** setting that corresponds to the circuit Bandwidth (BW). Therefore, it can be stated that the circuit bandwidth is **101.07673 MHz.**

Thus, in the given Bode plot (see Fig. 6.24), there is a low-frequency band, mid-frequency band, and high-frequency band. An amplifier gain is affected by the transistor's internal capacitances and externally connected capacitances in the circuit which reduce the transistor's gain in both low and high-frequency ranges of operation. A reduction of the circuit gain in a low-frequency band is due to externally connected coupling and bypass capacitors. They essentially behave like short circuits in mid- and high-frequency bands. A reduction of the circuit gain (and insertion of phase shift) in a high-frequency band is due to the presence of transistor (or any amplifying device) internal capacitances C_{be} and C_{bc} which become effective at high frequencies and are essentially open circuits in the low and mid-bands.

When a reactance of C_{be} becomes small enough, a significant amount of signal voltage is lost due to a voltage-divider effect of a source resistance and reactance of C_{be}. When a reactance of C_{bc} becomes small enough, a significant amount of an output signal voltage is fed back out of phase with input (negative feedback), and thus effectively reduces a voltage gain.

ix. Close the cursor readout display window. Now, change magnitude scale on the left-vertical axis (or y-axis) to linear (non-dB) to obtain the linear plot as shown in Fig. 6.25:

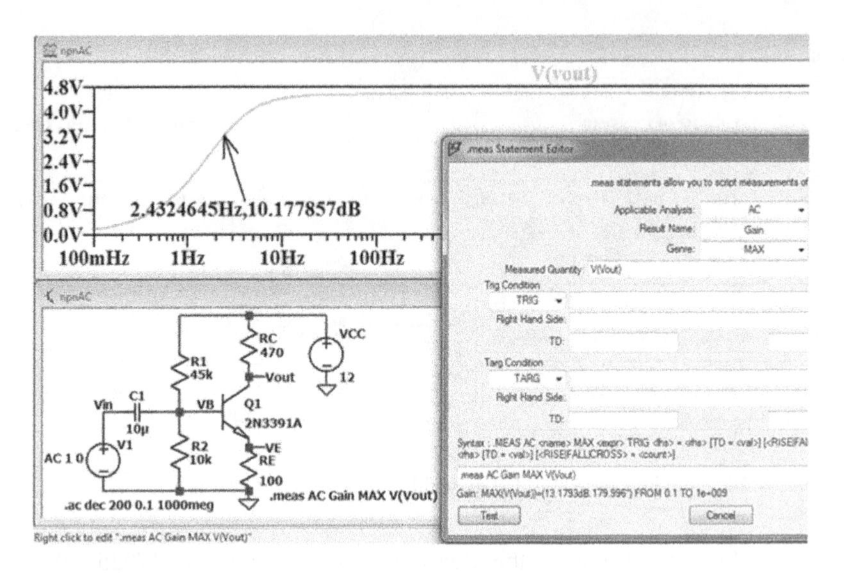

FIGURE 6.25 The Magnitude Plot on a Linear Scale.

x. Compute the circuit maximum gain using the .meas statement (see Fig. 6.25). It comes out to be 13.179256 dB or 4.560002 V.

6.3.2 Impedance/Resistance Computation

- **Plotting the Output Resistance:**

xi. Go to the View -> Visible Traces schematic icons. In the Select Visible Waveforms dialog window, press Alt + Double-Click on the selected trace name/title/label to open the Expression Editor so that the desired expression can be entered for plotting. As an output resistance of a transistor amplifier is given by output voltage at collector divided by collector current, enter an expression V (Vout)/Ic(Q1) in the Expression Editor window as shown in Fig. 6.26.

xii. Click the OK (first on the Expression Editor, and then on the Select Visible Waveforms window as shown in Fig. 6.26) to plot the circuit output resistance.

- **Editing with notes and annotations:** Open the Annotation Line/Color dialog window shown in Fig. 6.27 by right-clicking on the added notes and annotations (comments/shapes) to be edited in the plotted trace when a pointing finger appears.

Click the **Delete this Object** button for deletion. Here, the user can also change the line styles added on the trace (Fig. 6.27).

An output resistance plot of the transistor amplifier is shown in Fig. 6.28.

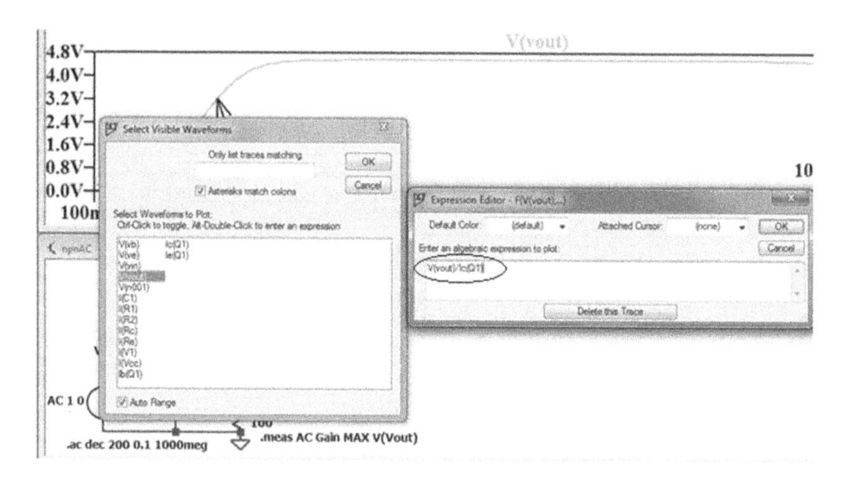

FIGURE 6.26 The Select Visible Waveforms Dialog Window.

FIGURE 6.27 The Annotation Line/Color Dialog Window.

- **Plotting the Resistance seen at the Base of the Transistor Model 2N3391A and Input Resistance:**

xiii. Right-click on the trace label to enter an expression V(VB)/IB(Q1) in the Expression Editor window for plotting the base resistance. Similarly, enter an expression V(Vin)/I(V1) in the Expression Editor window for plotting the input resistance.

A plot of **resistance** seen at the **Base** as well as the **Input Resistance** of the simulated amplifier using the transistor model **2N3391A** is shown in Fig. 6.29.

The resistance seen at the base is 32.9712 kΩ

Cursor 1
V(vb)/Ib(Q1)
Freq: 10KHz Mag: 32.971235KΩ ,

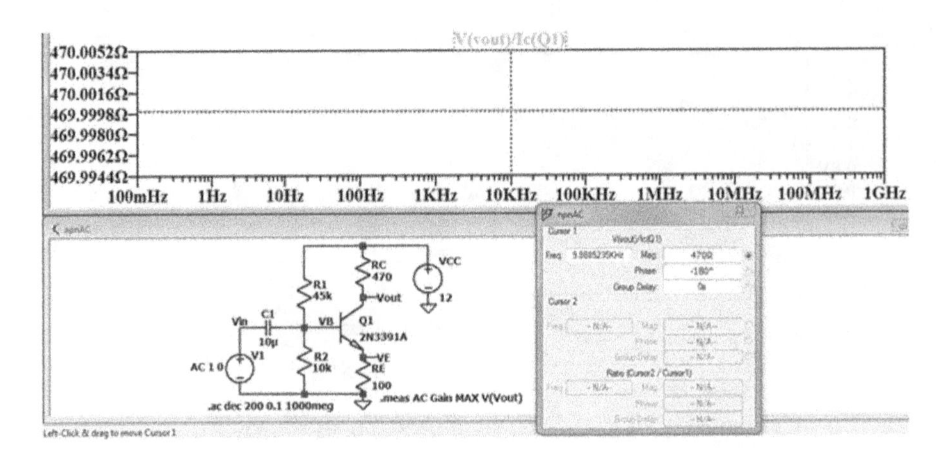

FIGURE 6.28 The Amplifier Output Resistance.

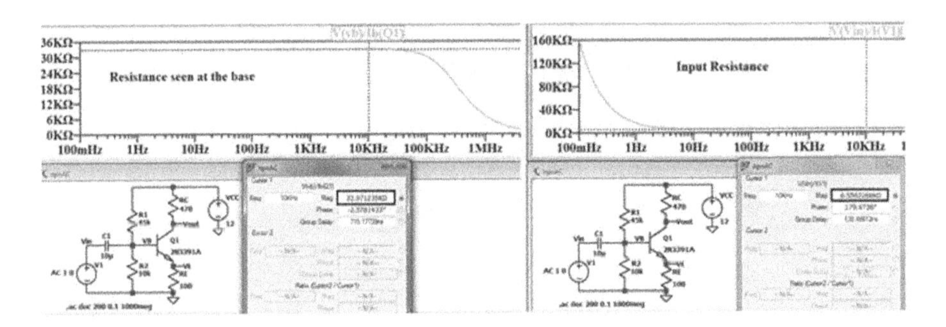

FIGURE 6.29 The Resistance at the Base as well as the Input Resistance.

whereas the input impedance is 6.5562 kΩ

Cursor 1
V(Vin)/I(V1)
Freq: 10KHz Mag: 6.5562268KΩ

Manually, the resistance seen at the base is $(1 + \beta)(r_e' + R_E) = (1 + 350)(2.0647 + 100) = 35722.645\ \Omega$. The value of beta is taken from the available manufacturer's sheet for the BJT model 2N3391A. The input resistance is $R_1 || R_2 || (1 + \beta)(r_e' + R_E) = 6657.083\ \Omega$.

6.3.3 Current Gain Computation

- **Finding beta (common-emitter current gain)**

xiv. Enter an expression Ic(Q1)/Ib(Q1) or IC(Q1)/IB(Q1) in the Expression Editor window to plot a common-emitter current gain (beta) of the

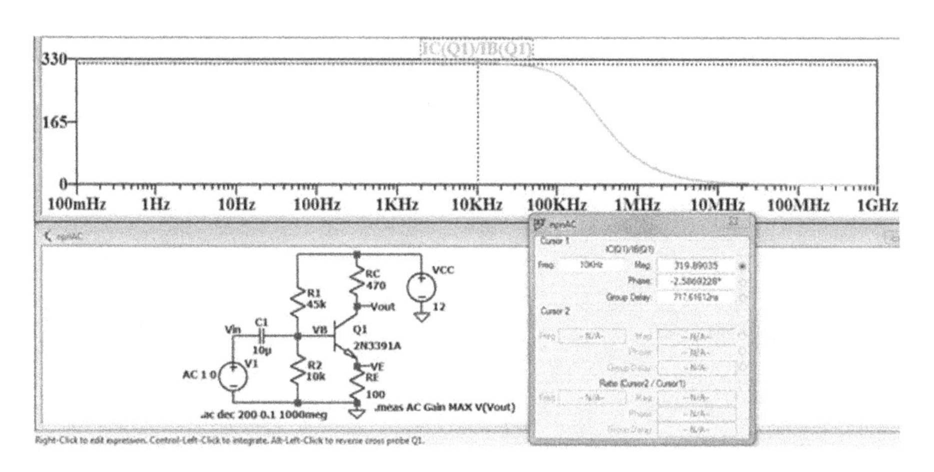

FIGURE 6.30 Magnitude Bode Plot of the Current Gain β.

simulated amplifier as shown in Fig. 6.30. The common-emitter current gain beta of an amplifier changes significantly at very high frequencies.

Because of a transistor base-emitter junction capacitance C_{be}, a transistor current gain β becomes frequency-dependent and can be represented as a single-pole function. For the earlier-drafted circuit, the DC current gain β_0 (current gain at 0 frequency) value is 319.9. For low frequencies, small-signal gain β does not vary too much from the DC gain value β_0 (Fig. 6.30).

6.3.4 Effects of Varying Emitter Resistance

- **Effects of Changing the Emitter Resistance Value on the Circuit Gain and Bandwidth**

xv. Open the property editor window of the resistor model R_E to edit the resistance parameter value to {R}. Place the directive statement .step param R list 100 200 on the circuit schematic. Re-run an AC analysis to obtain the following output. Attach the Cursor **1st & 2nd**. Navigate the cursors on the 2nd step simulation run which corresponds to the gain magnitude output for R_E = 200 Ω. Now, read a difference between the upper −3 dB point and the lower −3 dB point to compute the −3 dB bandwidth when emitter resistance R_E = 200 Ω.

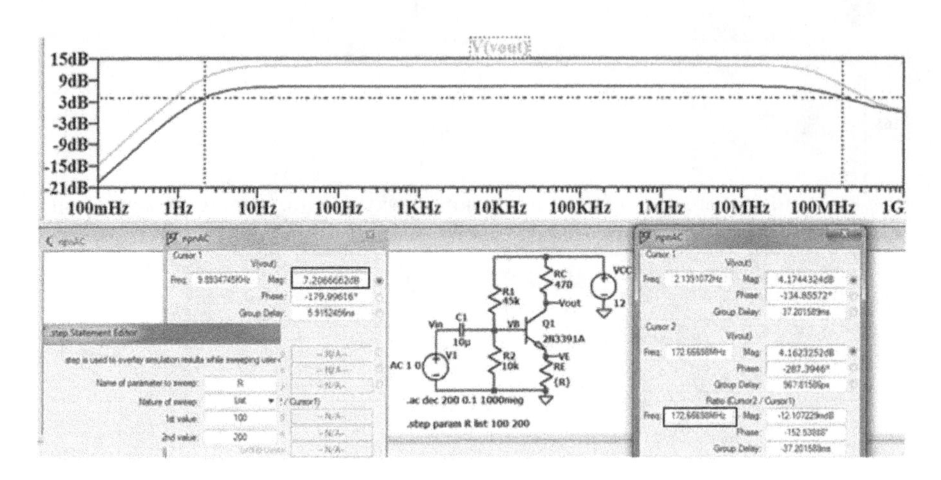

FIGURE 6.31 The Gain Magnitude Plot for Two Emitter Resistance Values.

From Fig. 6.31, it can be stated that as emitter resistance increases, gain decreases and the bandwidth increases. The circuit gain is **7.2 dB** and the bandwidth is **172.67 MHz** at R_E = 200 Ω.

7 Parametric Sweep Analysis

7.1 SWEEPING PARAMETERS (.param and .step)

A Parametric analysis allows simulation of an analysis (DC operating point, Transient, DC Sweep, and AC) while using a range of the component parameter/temperature values. It is important to plot voltages or currents by varying a parameter value of the component/device in the component attribute/value field for comparing the performance of the circuit. Instead of manually entering different values for a particular parameter of the component and then simulate the circuit for the changed values to analyze the responses, it is better to use the .step command to sweep across a range of values in a single simulation run and have a side-by-side comparison.

To set or change values of the component at once, and thereby avoiding editing individual component values every time or to define parameters in terms of other parameters/functions of other parameters, the .param directive is used. The parameters are then used to define the values of the components in their value fields.

- **How To Achieve Parametric Stepping:**
  1. Assign a variable name (parameter value identifier name) that should begin with an alphabetic character to the parameter value is needed to be stepped and enclose the assigned name in curly braces { }. The addition of the curly braces around the variable name is important to tell LTspice XVII that the variable is a parameter. An AC amplitude of the voltage/current source can also be stepped.
  2. Write the .step SPICE directive to describe how to step (or sweep) the particular parameter value. Add the .step command by selecting the SPICE Directive (.op) schematic menu that steps the parameter linearly by default. For example,

     .step param Amp 1 30 20

     implies a linear step of AMP (user-defined variable) from 1 to 30 with a step size of 20 so that the parameter value is 1 and 21.

The .param directive in the statement allows the simulator to run an analysis with user-defined variables in the Value fields of the components. Several .param statements can be made active in a simulation at the same time but the parameter value identifier names in each set must be distinctive. The .param directive statements can be continued to another line by using the + continuation character.

The .step SPICE directive command automatically varies the user-defined variables representing the parameter values (e.g., resistance of some resistor) of the components over some range or list of values and repeats a particular analysis chosen for simulation for the defined range or list of values. The command performs an analysis repeatedly for different values of a temperature, the model parameter (such as a resistive value or a transistor attribute), the global parameter, the independent source component, etc.
The variable steps for the parameter to be swept may be linear, logarithmic, or specified as a list of values.

- **Logarithmic Sweep by Octave:**

  .step oct param C 1uF 20uF 3

  means logarithmic sweep of capacitance value (named as the C) from 1 μF to 20 μF in 3 logarithmic steps per octave.
- **Logarithmic Sweep by Decade:**

  .step dec param R 1k 10Meg 2

  means logarithmic sweep of resistance value (named as the R) from 1 kiloohm to 10 Megaohms in 2 logarithmic steps per decade.
- **Sweep by Using a List of Values:**

  .step param R list 100 300 5k 90k

  LTspice XVII can step 3 parameters simultaneously. If the user sets up a default value of the component parameter defined by a variable name (e.g., .param C = 1uF), then there is no need to change the value of the component if it is not required to be stepped in a particular simulation.

7.2 SWEEPING RESISTANCE

The objective is to simulate a simple voltage divider circuit (having two resistances in series connected across a voltage source) using a DC operating point analysis while stepping/varying the lower resistor R2 parameter value from 10 ohms to 5 kiloohms in incremental steps of 100.

Set the parameter values of the components present in the drafted circuit (see Fig. 7.1) as follows:

i. A peak value of the input voltage source component is set to be 20 V and a resistance value of the resistor R1 is set to be 500 Ω.
ii. **Sweep the R2:** Instead of entering an absolute resistance (or a numeric value for the parameter resistance) in the attribute editor window of the resistor model R2 (lower resistor), enter the variable {R} in the Value field. Now, the R2 represents a resistance with the variable parameter value of R Ω.

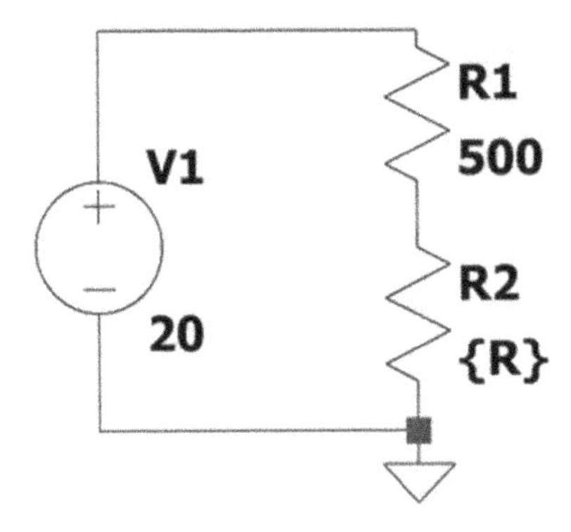

FIGURE 7.1 The Drafted Circuit (Voltage Divider).

Vary the parameter value R by using the .step directive: Click on the SPICE Directive (.op icon) schematic toolbar to open the **Edit Text on the Schematic** dialog window and type .step in a text box (ensure that the SPICE directive radio button is highlighted). After pressing the OK, place the directive near the circuit diagram. Now, right-click on the directive so that its syntax can be edited by using its LTspice XVII GUI (see Fig. 7.2).

In the **Name of parameter to sweep** field, enter the R (a user-defined name of the parameter to be swept). This defines the parameter R which can be swept by the .step directive. It is to be remembered that this name R is not a name of the component, instead, it is a name given by the user in the Value field of the resistor component's attribute editor window. With the SPICE directive statement, the value of R2 is increased from 10 Ω (start value) to 5 kΩ (stop value) in steps of 100 Ω as shown in Fig. 7.3.

Select the **Linear** option from the pull-down menu next to the **Nature of sweep** field. Click the OK to place the edited syntax of the complete .step directive statement automatically as shown in Fig. 7.4.

iii. Save the drafted circuit schematic file with a name, say parametric.asc.
iv. Now, specify a DC operating point analysis for simulating the drafted circuit named parametric (see Fig. 7.4).

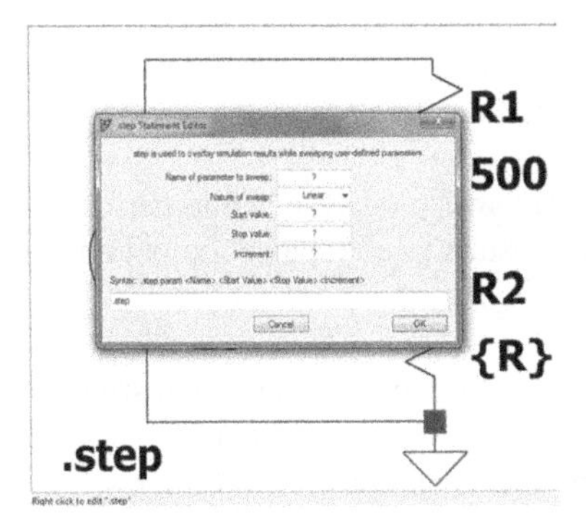

FIGURE 7.2 Editing the .step Statement.

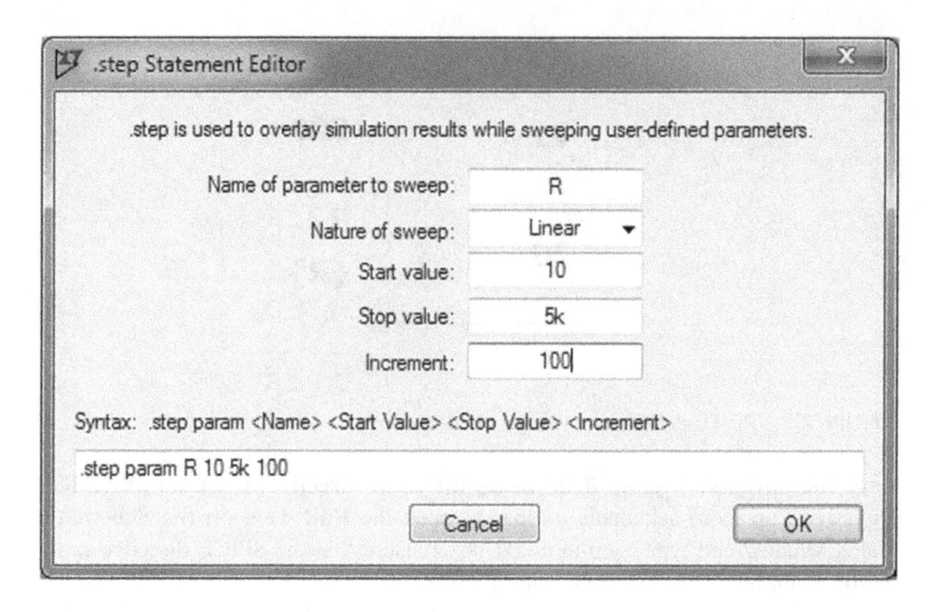

FIGURE 7.3 The complete.step Directive Statement.

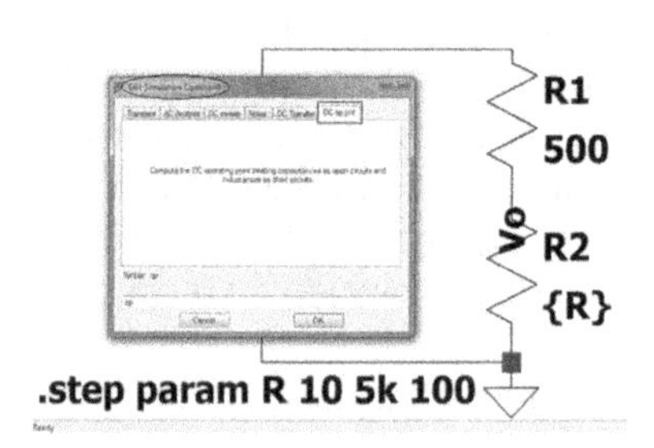

FIGURE 7.4 The .op Analysis with the .step Directive Statement.

Click the OK and place the .op directive somewhere on the schematic (Fig. 7.5).

v. Press the Run button (a running guy) to see the blank waveform window as shown in Fig. 7.5.

i. Probe a voltage at the node Vo to observe a voltage across the R2 for changes in the resistance values of the resistor R2 as shown in Fig. 7.6.

Remember: A parametric sweep analysis while performing a DC operating point (non-graphical analysis) results in a graphical output having the stepped parameter on a horizontal axis.

ii. It is also possible to sweep the parameter resistance using a list of values by selecting the **List** option from the pull-down menu next to the **Nature of sweep** field in the .step Statement Editor dialog window as shown in Fig. 7.7.

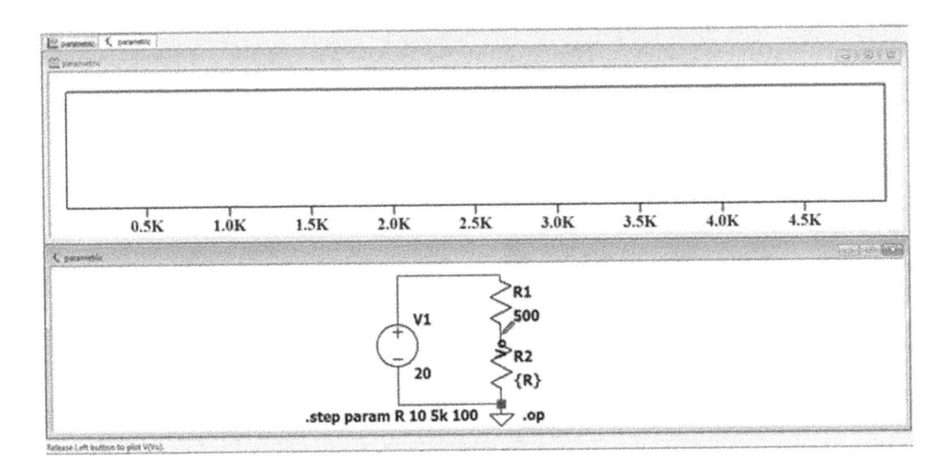

FIGURE 7.5 The Blank Waveform Viewer Window.

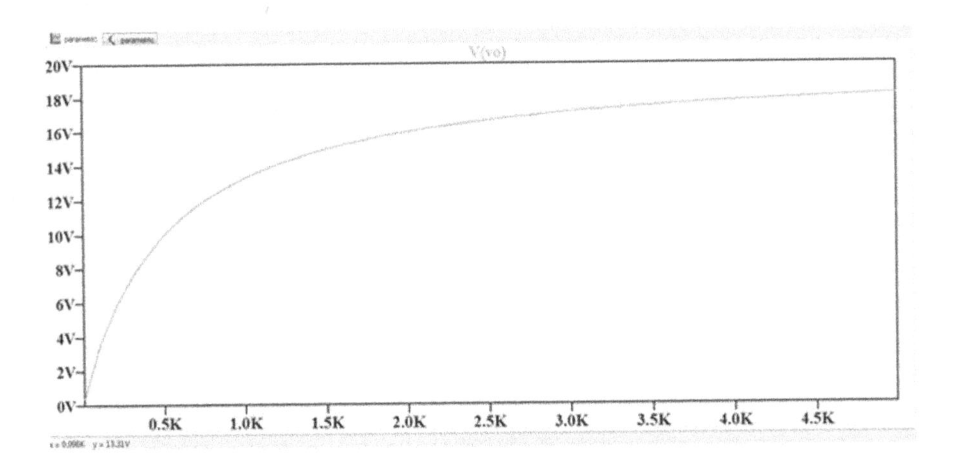

FIGURE 7.6 The Output Voltage for a Range of Resistance Values.

The Output Voltage for a List of the Resistor R2 Resistance Values, that is, 500 Ω, 1000 Ω, and 1500 Ω is shown in Fig. 7.8.

7.3 REPEATED AC ANALYSIS USING PARAMETER SWEEP

- **Repetitive AC analysis for a range of values of the R1**

In an AC analysis, the model parameters are usually fixed to calculate a small-signal AC response of a circuit. But, sometimes it is needed to view a response under varying parametric values of the components by stepping the parameter of interest using the .step command.

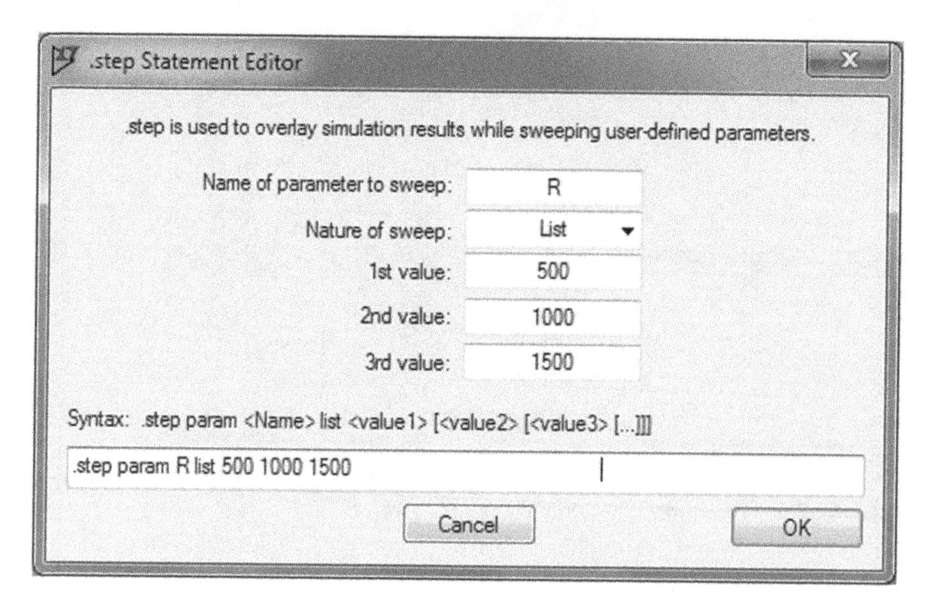

FIGURE 7.7 The .step Statement Editor.

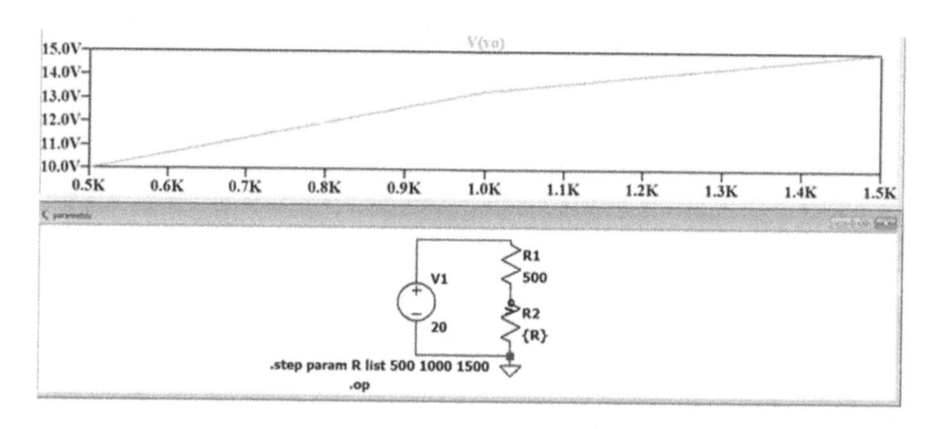

FIGURE 7.8 The Output for a List of Resistance Values.

For example, in the drafted RLC series circuit, it is required to sweep a resistance linearly through a range of 5 Ω to 30 Ω with an increment of 5 Ω using the .step directive (can also press an S key to insert the SPICE directive in the schematic editor).

i. To achieve the earlier-stated target, open the schematic file RLCSeries.asc. Now, edit the resistance parameter of the resistor R1 by typing {R} in place of entering a numeric value for the resistor model in the circuit schematic to sweep the parameter R. This defines R as a global parameter that can be swept using the .step param directive. Place the .step directive on the schematic with the syntax .step param R 5 30 5.

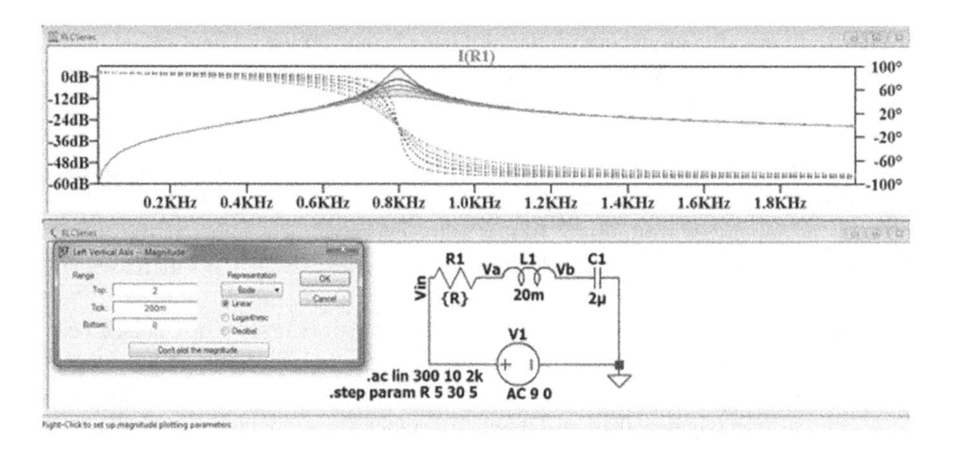

FIGURE 7.9 Repetitive AC Analysis for a Range of Values of the R1.

ii. Now, run an AC analysis simulation with the parameters settings as shown in Fig. 7.9 to obtain the following simulation results.

The Current Plots on a Decibel Scale (the default) against Variations in Frequency by Stepping the Resistance

iii. Right-click on the phase axis (i.e., on a right-vertical axis to set up phase/group delay plotting parameters) and select the **Don't plot the phase** option. Also, choose a representation of a Bode plot to be linear to obtain the circuit current magnitude on a linear scale as shown in Fig. 7.10.

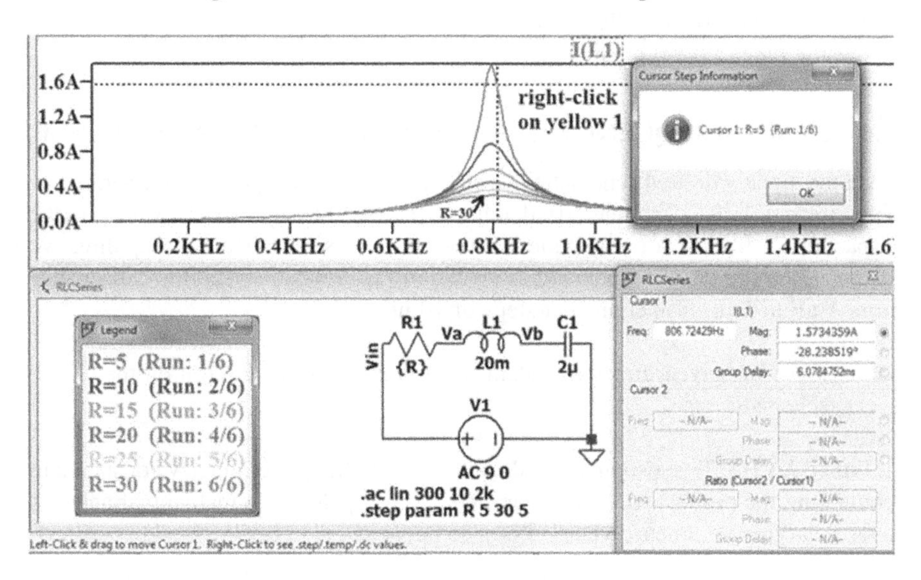

FIGURE 7.10 The Legend and Cursor Step Information Windows.

iv. Right-clicking anywhere on the plot pane brings up the context-based menu. Now, select the **View** menu, and then the **Step Legend** sub-menu for displaying legends.

The series RLC circuit current plots on a linear scale against variations in frequency for different values of the resistance parameter are shown in Fig. 7.10.

It can be concluded from Fig. 7.10 that as the parameter R is reduced, the current waveform becomes sharper and selectivity of the series RLC circuit increases.

v. In parametric sweep analysis, the numbered cursor (Cursor 1 or Cursor 2) can also be employed to know which trace belongs to which run of the .step in combination with .ac/.dc/.temp set of simulation runs. The attached cursor can be easily navigated from a dataset to another dataset with an up/down keyboard cursor key, and then right-click on the cursor (when yellow 1 or yellow 2 representing the cursors appears) to see the step information for that run.

Remember: Parametric analysis while performing an AC analysis (or any other graphical analysis such as transient and so on) results in a graphical output with a series of lines, one for each value of the stepped parameter.

vi. Now, to isolate one trace, use the command **Select Steps** from the trace menu **Plot Settings** obtained by left-clicking anywhere on the waveform viewer window.

(**Note:** Use of the .step command with an AC analysis can drastically increase a simulation time, so carefully choose start and stop values, ranges, increments, and frequency range for each parameter sweep.)

7.4 SINGLE FREQUENCY ANALYSIS WITH SWEPT CAPACITANCE

Load the already drafted schematic by selecting the File -> Open -> RLCSeries.asc. To set the C1 as a parameterized object, change the capacitance value of the Capacitor C1 from C (or any numeric value) to **{C}**. Now, place the .step directive statement on the circuit schematic as shown in Fig. 7.11 to sweep the capacitance from 1 nF to 2 μF in incremental steps of 10 nF.

i. Place the given directive statement .step param C 1n 2u 10n on the circuit schematic.

LTspice XVII offers a sophisticated solution for holding a frequency constant and comparing the performance of the drafted circuit by doing a small-signal analysis over a varying parametric value of any component.

ii. For specifying a single frequency, change the simulation sweeping type from a decade to list. It is accomplished by selecting the **List** option from the pull-

.step Statement Editor

.step is used to overlay simulation results while sweeping user-defined parameters.

Name of parameter to sweep: C

Nature of sweep: Linear

Start value: 1n

Stop value: 2u

Increment: 10n

Syntax: .step param <Name> <Start Value> <Stop Value> <Increment>

.step param C 1n 2u 10n

Cancel OK

FIGURE 7.11 Sweeping the User-Defined Parameter C.

down menu next to the **Type of sweep** field while doing an AC analysis as shown in Fig. 7.12.

iii. Specify a simulation frequency (single frequency at which an AC analysis is needed to be performed) as 1k (equal to 1 kHz) in the respective **Edit Simulation Command** dialog window under the **AC Analysis** tab as shown in Fig. 7.12.

iv. Click the OK and place the .ac statement on the circuit schematic (see Fig. 7.13) for doing a single frequency analysis with swept capacitance.

v. Save the file as singlefreq with an extension .asc.

vi. Run the simulation. Left-click directly on the component L1 to plot a current

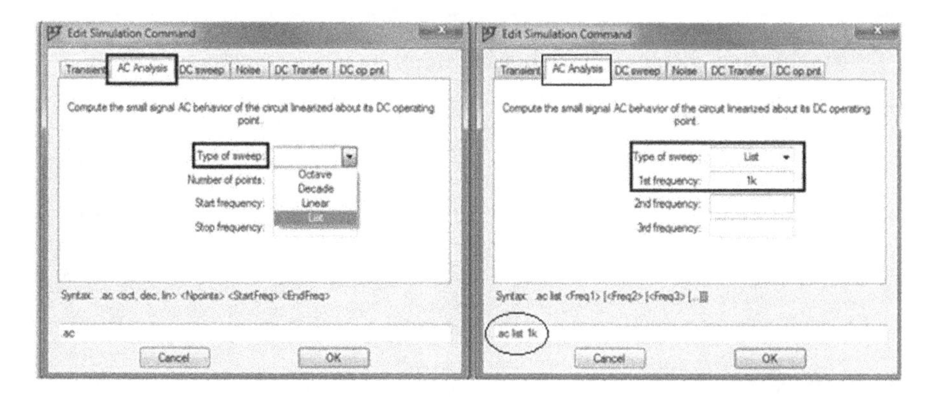

FIGURE 7.12 Setting the AC Analysis Parameters.

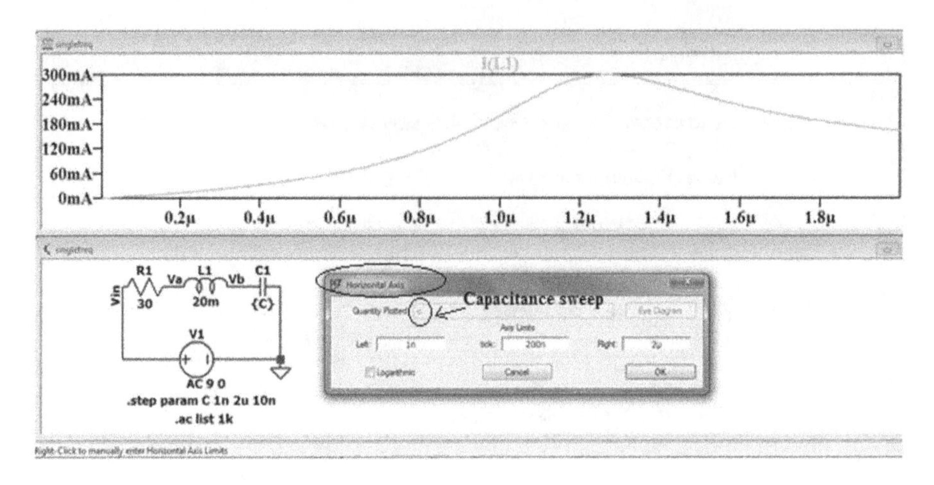

FIGURE 7.13 An AC Simulation while Keeping Frequency Constant.

through the component (here, the series circuit current) as a function of capacitance (see Fig. 7.13). In a single frequency analysis with the .step command, the resulting plot shows magnitude and phase as a function of the parameter swept i.e. the capacitance and not the frequency.

From Fig. 7.13, it can be seen that the result of an AC simulation with a single frequency uses a capacitance sweep (defined in the .step directive command) as an x-axis.

(**Note:** An x-axis is a capacitance as defined by the .step command)

vii. Use the .meas directive to find when the maximum circuit current flows at a fixed frequency of 1 kHz and it comes out to be −10.4588 dB or 299.958 mA (see Fig. 7.14). Now, attach the Cursor 1 to read a capacitance value in the

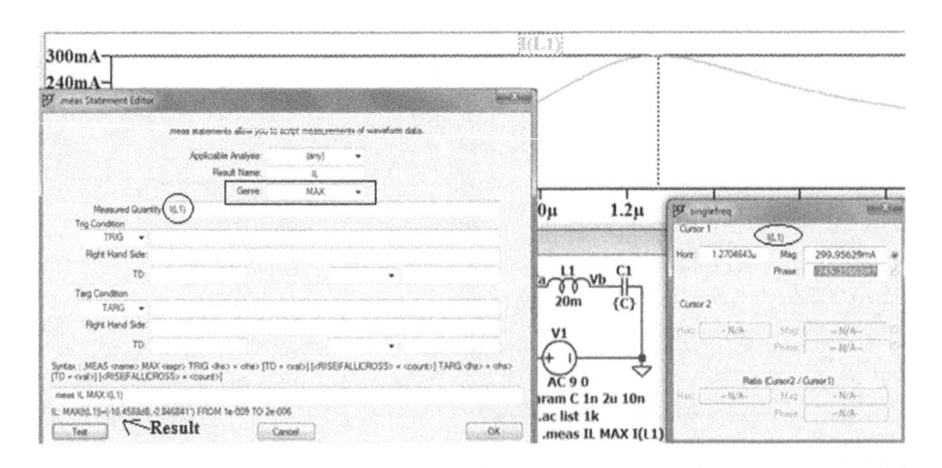

FIGURE 7.14 The Numbered Cursor for a Capacitance at a Specified Current.

Horz box at the maximum current level of 299.958 mA in the Mag box as shown in Fig. 7.14. Thus, the simulation results show that a maximum current flows through the circuit at C = 1.27 μF.

Manually, at a frequency of 1 kHz (corresponds to a resonant frequency when the circuit current is maximum) the value of capacitance should be equal to:

$$C = \frac{1}{4\pi^2 f^2 L} = \frac{1}{4\pi^2 (1000)^2 (20 \times 10^{-3})} = 1.2665\ \mu\text{F}$$

Thus, it can be stated that the theoretical value agrees with the experimentally observed value.

Therefore, C = 1.2686585 μF (microfarads) for a resonance to occur at 1 kilohertz.

7.5 REPEATED OPERATING POINT ANALYSIS ON TRANSISTOR

- **Finding Resistance for the Specified Transistor Operating Point Voltage**
  i. Draw the circuit schematic as shown in Fig. 7.15 and save it as transistorsweeptf.asc. To keep the selected transistor model 2N3391A in an active region, the circuit needs to be biased properly so that the Vout is equal to $V_{CC}/2 = 6$ V when V_{CC} is taken to be 12 V. It is desired to find a value of the R1 to fix the transistor operating point voltage (i.e., collector-emitter voltage) at 6 V. A collector-emitter voltage is equal to a difference between collector voltage and emitter voltage.
  ii. Sweep the parameter R by using the directive statement

.step param R 5k 200k 10k.

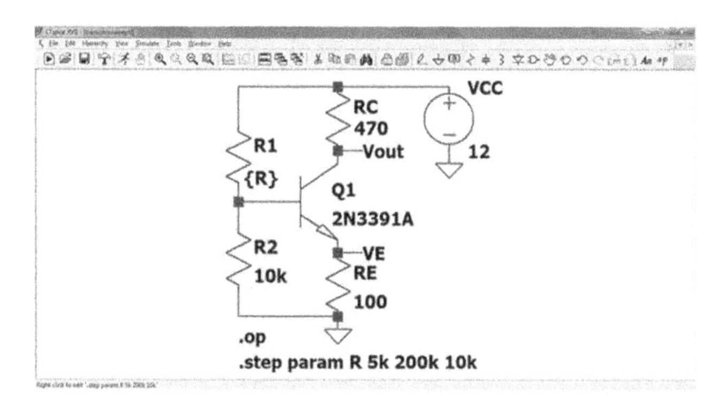

FIGURE 7.15 The Circuit Schematic.

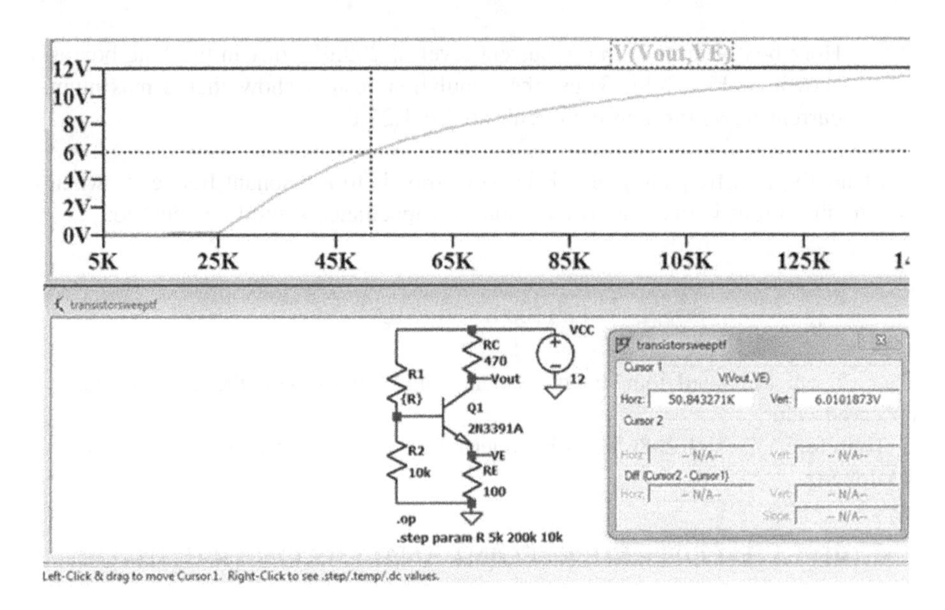

FIGURE 7.16 The Collector-Emitter Voltage against Variations in the Parameter R.

iii. Run the .op analysis statement. Probe the differential voltage between collector and emitter (i.e. a voltage at the node Vout with reference to the node VE) to obtain the following output:

From Fig. 7.16, in the cursor window, the resistance value of the R1 can be read from the Cursor 1 **Horz** box which is equal to 50.8 kΩ.

iv. Probe the collector current IC(Q1) versus the bias resistance R1 to obtain the following output as shown in Fig. 7.17:

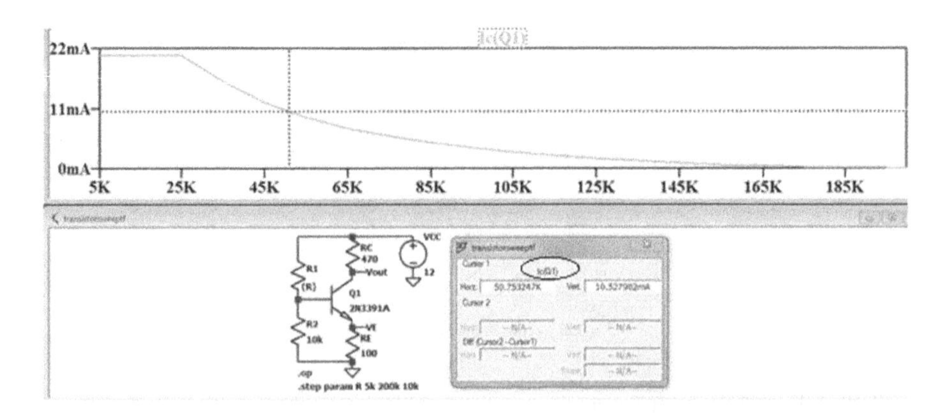

FIGURE 7.17 The Collector Current against Variations in the Parameter R.

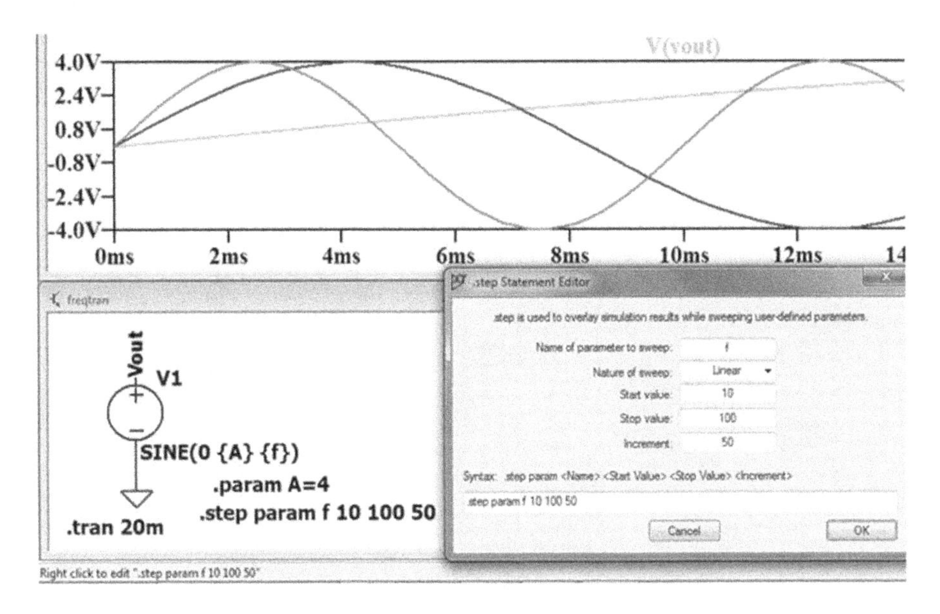

FIGURE 7.18 A Transient Analysis Simulation Output with a Varying/Swept Frequency.

At R1 = 50.8 kΩ, the collector current comes out to be 10.528 mA (after rounding). As the computed current value 10.528 mA is near to the value required to bias the transistor at a mid-point on the load line i.e. $V_{CC}/2R_C = 12/(2 \times 470) = 12.67$ mA. Therefore, the transistor remains in an active region.

7.6 SWEEPING FREQUENCY FOR TRANSIENT SIMULATIONS

The target is to vary/sweep a frequency of the independent voltage source component V1 (set as a sine wave signal generator under the Functions setting) along time so that the output of the system can be seen in a time-domain when the frequency changes. Draw the circuit schematic and save it. Parameterize the voltage source V1 amplitude as well as frequency as {variable_names}. Using the .param command, assign a value to the amplitude parameter. Using the .step param command, sweep the frequency parameter from start value of 10 Hz to stop value of 100 Hz with an increment of 50 Hz. Run the .tran simulation with a varying frequency to obtain the output as shown in Fig. 7.18.

If the frequency is too less so that spacing/time between the samples is very large, the output may not be a proper sine wave.

8 DC Transfer Analysis

8.1 EQUIVALENT CIRCUIT AND DC TRANSFER SIMULATION

A voltage divider consists of two resistances R1 and R2 connected in series across a supply voltage Vs. The supply voltage is divided between the two resistances to give an output voltage Vo across the R2 (connected between the terminals say A and B). An output of any circuit (or device) is equivalent to an output impedance (or resistance) in series with an open-circuit voltage and is called an **equivalent circuit of the original one** (Fig. 8.1).

a. **Output Voltage V_o**
   A value of an open-circuit voltage (Thevenin's voltage) in an equivalent circuit is the output voltage V_o between the A and B when a load is disconnected from the output terminals A and B and can be computed using Thevenin's equivalent circuit method.
b. **Output Impedance (or Resistance) of a Voltage Divider**
   An output impedance (a term) can be thought of as being an impedance (or resistance) that a load perceives when looking back into the terminals across which a load is connected with an input voltage equal to zero (i.e., the input V1 is short-circuited). In the equivalent circuit of a voltage divider, an output impedance (Thevenin's resistance) is just the resistance R_{out} that is equal to two resistances R1 and R2 connected in parallel, that is,

$$R_{out} = \frac{R1 \times R2}{(R1 + R2)} \tag{8.1}$$

For practical use, an output impedance of a voltage divider should be much smaller (**less than a tenth of an input impedance**) than an input impedance of a load circuit connected to it.

c. **Input Impedance (or Resistance) of a Voltage Divider**
   An input impedance is defined as a total sum of resistance, inductance, and capacitance (if present in the circuit) seen by an input source.

8.1.1 Simulating Voltage Divider Using DC Transfer Analysis

- **Steps Involved:**
  i. Construct the schematic diagram having a series combination of the R1 and R2 across the V1 for a DC transfer analysis. Set the parameter values

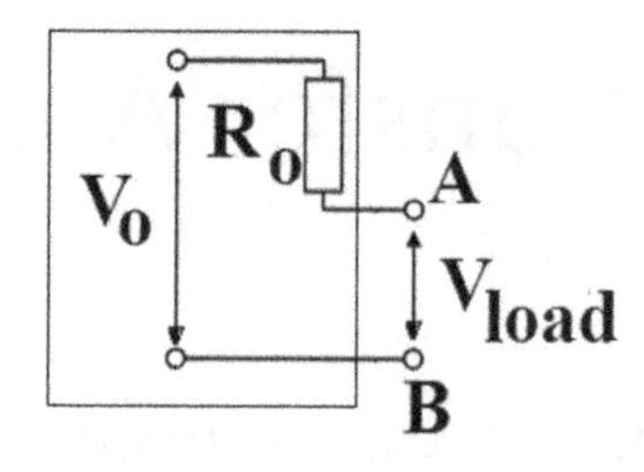

FIGURE 8.1 A Thevenin Equivalent Circuit of a Voltage Divider.

of the components V1, R1, and R2. Label the output node as OUT using the Label Net tool. Save the file.

ii. Right-click on the body of the V1 to open the component menu window. Click on the **Advanced** button and select the **(none)** radio button under the Functions setting. Afterward, click the OK. Now, ensure that the value **V** (or any other parametric value if prescribed earlier) described near the signal source V1 is no longer visible as shown in Fig. 8.2.

iii. Select the Simulate -> Edit Simulation Cmd schematic options to open the Edit Simulation Command dialog window. Here, choose the **DC Transfer** tab by clicking on it. Set the respective fields as shown (enter the V(OUT) in the box next to the Output field and the V1 in the text box next to the Source field):

In the **DC Transfer** Tab Dialog Window, under the **Find the DC small-signal transfer function** setting, enter:

Output: V(OUT)
Source: V1

Click the OK to place the statement **.tf V(OUT) V1** on the schematic.

After placing the .tf directive on the schematic, press the Run schematic tool button. The Transfer Function floating little window pops up. A list of the simulated parameters obtained after running a DC Transfer analysis can be seen in Fig. 8.2.

The following results are obtained after running the simulation:

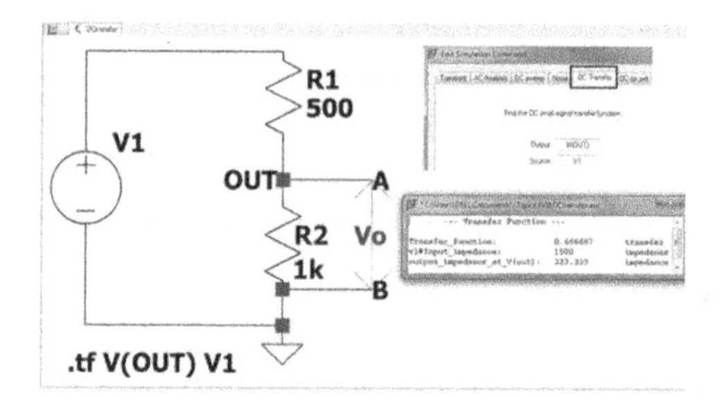

FIGURE 8.2 The Simulated Output.

The Simulation Results:

```
--- Transfer Function ---

Transfer_function:                 0.666667    transfer
v1#Input_impedance:                1500        impedance
output_impedance_at_V(out): 333.333     impedance
```

From the simulation results, it can be stated that the open circuit output voltage (Thevenin's voltage for an equivalent circuit) between terminals A and B (when R_L is disconnected) is equal to the circuit transfer function multiplied by the input, that is, i.e. 0.666667 × 20 = 13.33334 V.

$$R_o = \frac{500 \times 1000}{(500 + 1000)} = 333.3333333\ \Omega = R_{th} \tag{8.2}$$

iv. Theoretically, the circuit **output impedance/resistance** (Thevenin's equivalent circuit resistance) for an equivalent circuit representing the original one is equal to:

The circuit **input resistance** is equal to:

$$R_{in} = R_1 + R_2 = 1500\ \Omega = equivalent\ \ resistance \tag{8.3}$$

The circuit transfer function/gain is equal to:

$$gain = \frac{V_o}{V_1} = \frac{V_1 R_2}{V_1(R_1 + R_2)} = \frac{1000}{(500 + 1000)}$$
$$= 0.6666666667 = transfer\ \ function \tag{8.4}$$

Because the output voltage across the R2 = $V_o = \dfrac{V_1 R_2}{(R_1 + R_2)}$ (8.5)

Thus, it can be stated that the obtained results of the simulation are the same as theoretical results (Fig. 8.3).

8.1.2 Equivalent Resistance of Network

i. Draw the schematic and perform a DC transfer analysis on the circuit diagram shown in Fig. 8.3.
ii. Right-click on the body of the V1 and select (none) radio button under the Functions setting. Click the OK. Ensure that the value **V** (or any other parametric value if prescribed earlier) described near the signal source V1 is disappeared.
i. Run a DC transfer analysis to obtain the following output:
ii. Manually, the circuit equivalent resistance is:

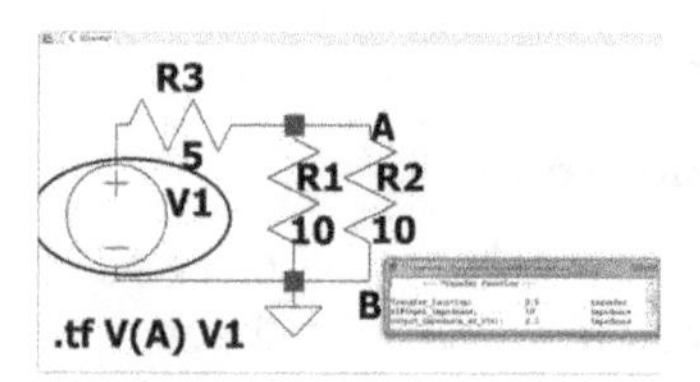

FIGURE 8.3 Computing the Equivalent Resistance and Transfer Function.

$$R_{eq} = \frac{10 \times 10}{20} + 5 = 10\ \Omega \tag{8.6}$$

Thus, both the simulation result (input impedance) and the theoretical result (equivalent resistance) agree.

8.2 THE .tf (DC TRANSFER) ANALYSIS ON BJT

i. Open the schematic file named npntransient.asc (see Section 5.4) and run the .op analysis to find the operating base voltage needed to forward-bias the base-emitter junction of the designed voltage-divider biased amplifier circuit (see Fig. 5.23). The operating point voltage at the base is observed as 1.92497 V.
ii. Now, draw the new circuit schematic and perform a DC transfer analysis on the circuit diagram to find the resistance seen at the base terminal as shown in Fig. 8.4.

The Simulated Output:

```
--- Transfer Function ---

Transfer_function:                  -4.55999  transfer
vb#Input_impedance:                 33004.8   impedance
output_impedance_at_V(vout): 468.367   impedance
```

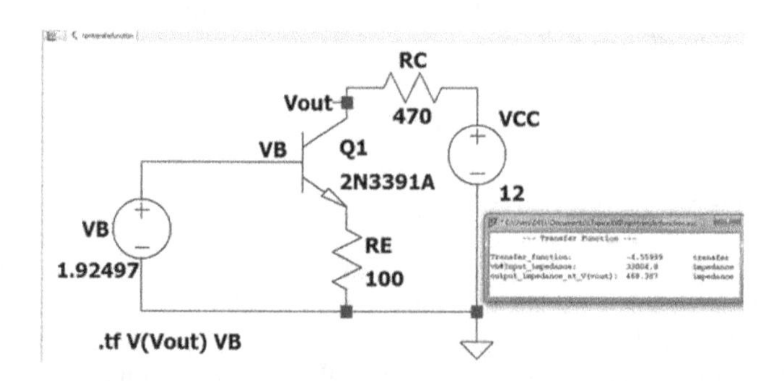

FIGURE 8.4 The Resistance Seen at the Base of the Voltage-Divider Biased Amplifier.

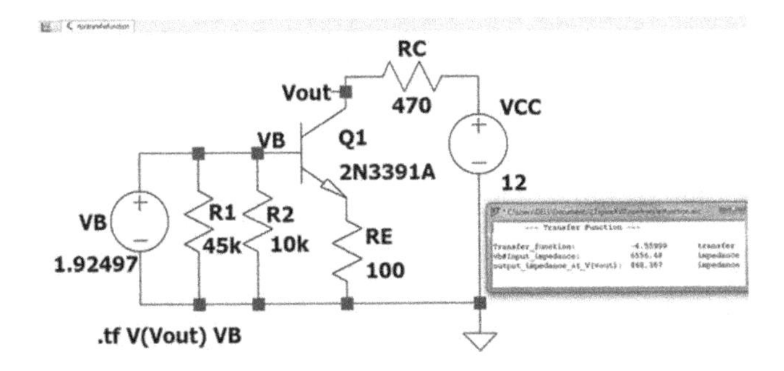

FIGURE 8.5 The Input Resistance of the Voltage-Divider Biased Amplifier.

Thus, the resistance seen at the base terminal is 3.3004 kΩ. A voltage gain for the above-designed voltage-divider biased amplifier comes out to be –4.55999. The output impedance is 468.367 Ω.

iii. Re-draw the schematic and perform a DC transfer analysis on the circuit diagram to find the input resistance as shown in Fig. 8.5.

The Simulated Output:

```
--- Transfer Function ---

Transfer_function:                -4.55999 transfer
vb#Input_impedance:               6556.48  impedance
output_impedance_at_V(vout):      468.367  impedance
```

Thus, the input resistance is 6556.48 Ω. A voltage gain for the above-designed voltage-divider biased amplifier comes out to be –4.55999. The output impedance is 468.367 Ω.

9 Small Projects (Examples)

9.1 FULL-WAVE BRIDGE RECTIFIER WITH FILTER

A bridge-wave rectifier is a full-wave rectifier that uses four diodes in a bridge circuit configuration to efficiently convert an AC into a unidirectional current (pulsating DC). A **capacitor** connected across a rectified output allows an AC signal to pass through it and blocks a DC signal, thereby acting as a high-pass filter. Thus, AC ripples in a DC output voltage get bypassed through a parallel **capacitor** circuit, and a pure DC voltage is obtained across a load resistor.

- **Simulation of a full-wave bridge rectifier with a smoothing capacitor (filter):**
  i. Draw the circuit schematic as shown in Fig. 9.1.
  ii. Select the SINE function for the voltage source V1, and then fill in the information as follows:

  DC offset: 0
  Amplitude: 10
  Freq (Hz): 50
  Leave all the other boxes blank.

  iii. Pick the 1N914 diode model as it is capable of handling large currents and voltages in a bridge-wave rectifier. To change the rectified voltage from a pulsating DC voltage to a steady one, place a capacitor across the load. The capacitor charges during the positive-going portion of each half-cycle, and it discharges through the load during the negative-going part. Depending on an RC time constant the discharge takes more or less of the time between positive half-cycles, and the filtering action flattens the output voltage to a greater or lesser extent. A time constant should be large compared to a period of the sinusoidal input.
  iv. Type {C} in the value field of the C1 and 100k (stands for 100 kiloohms) in the value field of the R1.
  v. Place the .step command with the following syntax:

  .step param C list 10n 100n 2u

  vi. Run a transient simulation for 60 milliseconds. As the period of an input sine wave applied is T = 1/50 sec, to cover 3 periods choose a simulation time to be 60 ms.

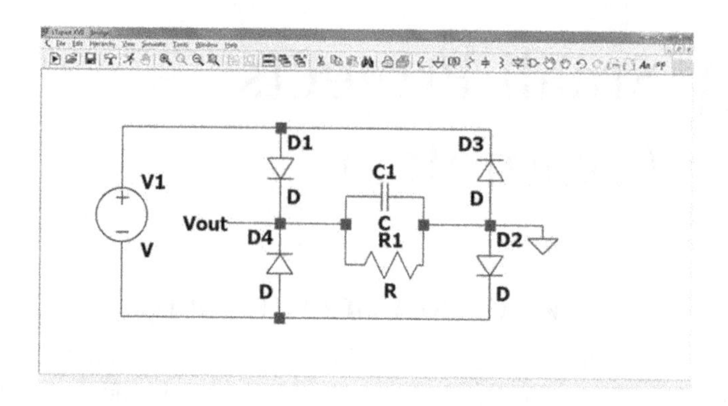

FIGURE 9.1 The Circuit Schematic.

The output (Fig. 9.2):

Because of the junction potential of the two diodes that are in series with the load, the amplitude of the output DC voltage lower than 10 V by about 1 V. This difference represents the junction potential.

Effects of changing a capacitance value on the rectified output: It can be observed that capacitive filtering afforded by the 10 nF capacitor raises the tail of a half-cycle, the 100 nF capacitor smoothens the waveform even more but still have a large ripple and the 2 μF capacitor takes out enough of the ripple to make the output DC voltage essentially flat. Thus, it can be concluded that the larger an RC time constant the flatter the exponential decay, and the closer the approximation to a constant waveform.

Remember: For practical purposes, there are some constraints on increasing a capacitance value (to reduce ripples) which need to be followed for a safer operation in real circuits. The maximum current ratings of diodes used in the design of any circuit must be considered.

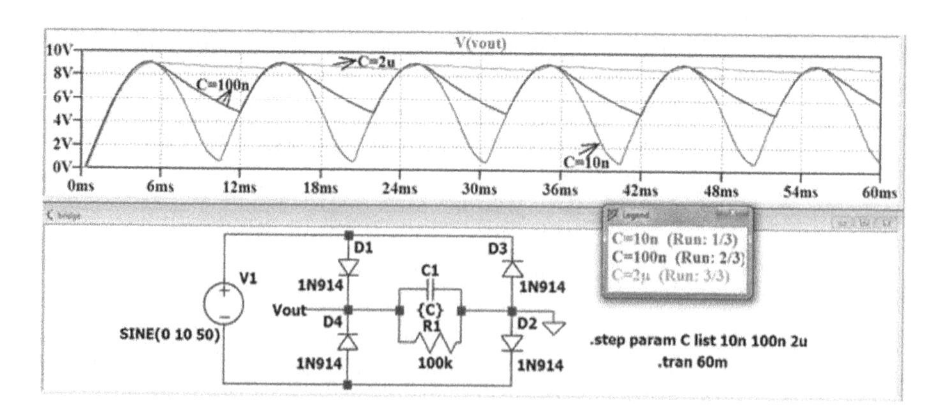

FIGURE 9.2 The Filtered Output of a Bridge-Wave Rectifier.

(**Note:** It is to be noted that all the diode models do not follow a 0.7 V drop assumption and may switch on and off more quickly, for example, the 1N914 signal diode and BAT85 Schottky diode.)

Also, instead of changing the parameter value (capacitance) of the C1, the user can vary the frequency of the input signal applied to observe its effect on the ripples.

Important: The diodes used in a rectifier circuit must adhere to the proper ratings (the maximum average forward current and peak inverse voltage) which they can withstand without damage.

9.2 OP-AMP AS INVERTING AMPLIFIER

Operational amplifiers (OP-Amps) are linear devices designed for acquiring voltage amplification, signal conditioning, filtering, or for performing various mathematical operations such as addition, subtraction, integration, differentiation, etc. They are driven by external DC voltage supplies (+Vs, −Vs) which are ordinarily symmetrical and generally taken in a range of ± 10 to ± 20 V and the output of an OP-Amp is limited to approximately 1.5 V less than these supply voltages. A limiting output implies that an OP-Amp can only amplify signals to a level within a range of supply voltages used, and therefore it is impossible for an OP-Amp to generate a voltage $> +Vs$ or $< -Vs$. A constraint i.e. $-Vs < vout < +Vs$ may result in clipping of peaks of an output sine wave and can produce distortion in an output wave.

i. **To Run a transient analysis for the drafted circuit schematic (see Fig. 9.3):**
   i. Construct the circuit schematic of an OP-Amp as an inverting amplifier with a gain value ($A_v = -R_F/R_1$) of −10.
   ii. Note down the port name and connections to create the positive and negative DC source voltages for the Op-Amp and location of the ground nodes.
   iii. Set the independent voltage source component V3 connected to the negative input terminal of the OP-Amp OP747 to be a sine function of peak amplitude value 1 V and frequency 100 Hz.
   iv. Run a transient simulation for 0.02 s (2 cycles of the input sine wave).

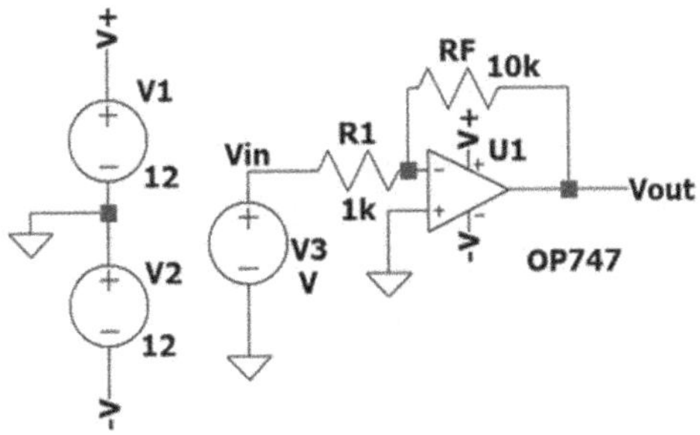

FIGURE 9.3 Simulating the OP-Amp as an Inverting Amplifier.

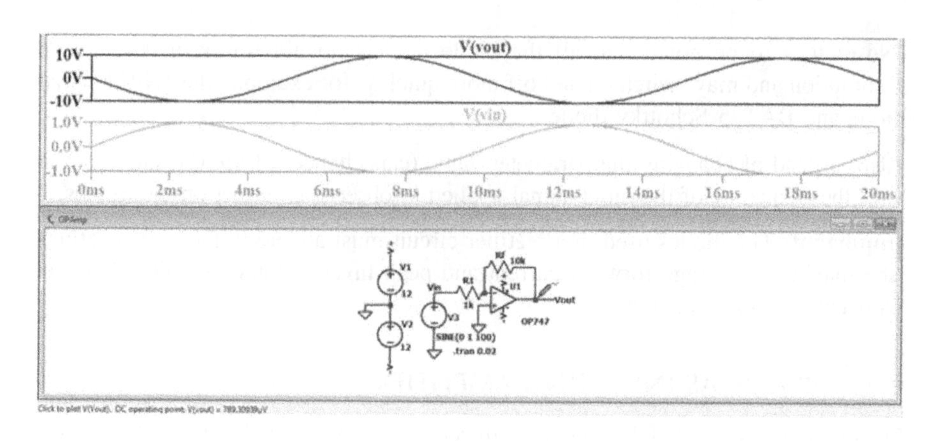

FIGURE 9.4 The Inverting Amplifier Output.

Case (a): The output when an input sine wave is having a peak amplitude of 1 V (Fig. 9.4)

v. Now, set the independent voltage source component V3 connected to the negative input terminal of the OP-Amp OP747 to be a sine function of peak amplitude 5 V and frequency 100 Hz.

Case (b): The output when an input sine wave is having a peak amplitude of 5 V (Fig. 9.5)

vi. Right-click on the SINE(0 5 100) and delete the information in a text box so that the parameter function value SINE(0 5 100) near the V3 disappears. Change the name of the V3 to Vinput and connect it to the positive terminal of the OP-Amp OP747. Connect the negative terminal of the source to the

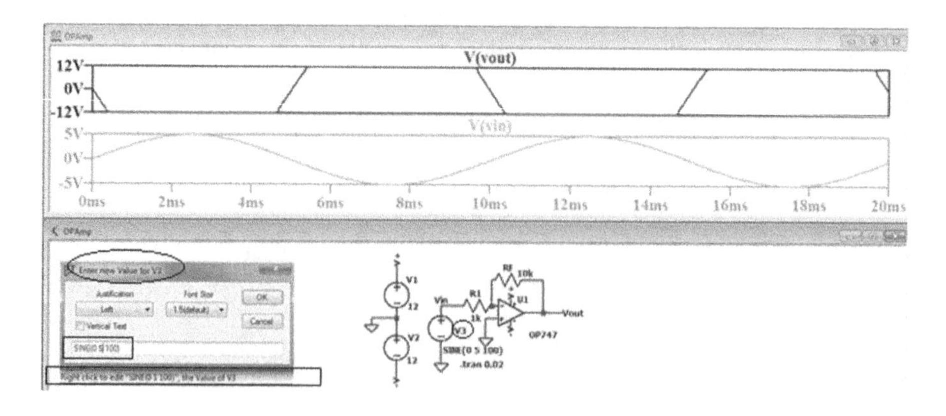

FIGURE 9.5 The Clipped Output of an Inverting Amplifier.

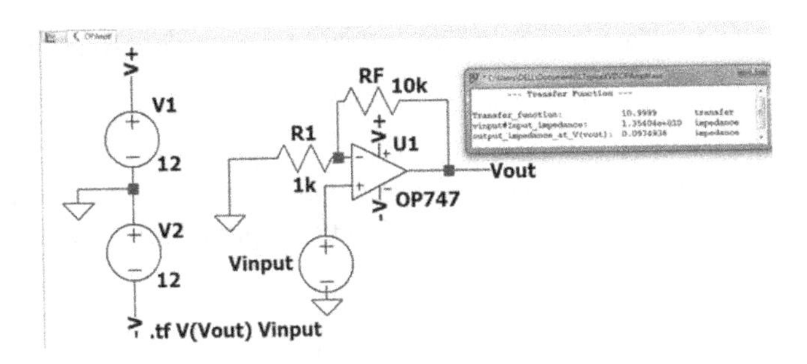

FIGURE 9.6 The Non-Inverting Amplifier Output.

ground as shown in Fig. 9.6 and save the schematic with the name OPAmptf. This results in a non-inverting amplifier (gain $A_v = (1 + R_F/R_1)$ using an OP-Amp.

vii. Perform a DC transfer analysis (.tf) on the circuit schematic in Fig. 9.6 by placing the **.tf V(Vout) Vinput** analysis statement on the schematic to see the simulated results:

```
--- Transfer Function ---

Transfer_function:                10.9999       transfer
vinput#Input_impedance:           1.35404e+010  impedance
output_impedance_at_V(vout):      0.0974936     impedance
```

9.3 ACTIVE FILTER USING OP-AMP

Operational amplifiers (OP-Amps) can be combined with energy-storage elements to design active filters for generating frequency-dependent responses.

i. Draw the schematic diagram of a single-pole low-pass active filter as shown in Fig. 9.7 by connecting the capacitor component C1 of value 2nF in parallel

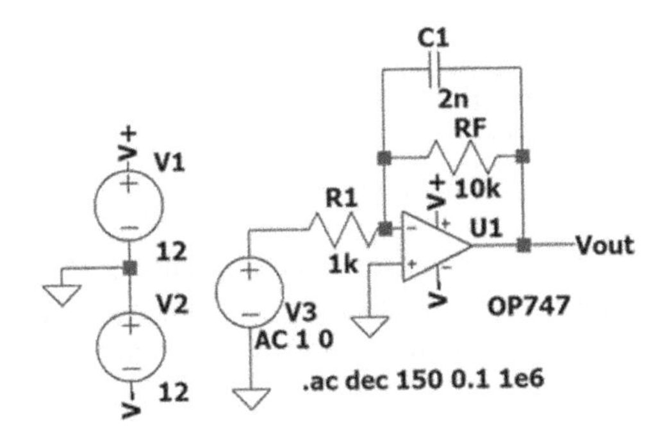

FIGURE 9.7 The OP-Amp as a Filter in an Inverting Configuration.

with the feedback resistor $R_F = 10\ \text{k}\Omega$ using an inverting amplifier configuration.

A cut-off frequency for a gain magnitude versus the frequency response of an amplifier is defined as a point where the gain magnitude starts to decrease with frequency. The relationship between a cut-off frequency, R_F and C_1 is given by Eq. (9.1):

$$f_c = \frac{1}{2\pi R_F C_1} = \frac{1}{(2 \times 3.14 \times 10000 \times 2 \times 10^{-9})}$$
$$= 7961.78343949\ \text{Hz} = 7.96\ \text{kHz} \tag{9.1}$$

Therefore, the cut-off frequency f_c (where the magnitude response reduces by −3 dB from its maximum value) for the earlier circuit in Fig. 9.7 should be equal to 7.96 kHz.

ii. The target is to apply an AC analysis (frequency sweep) simulation with the independent voltage source changed to an AC version with a 1 V amplitude.

To set an AC analysis simulation and sweeping frequency from 0.1 Hz to 1 MHz, do as follows:

- On the schematic menu, left-click the Simulate -> Edit Simulation Command options on the schematic menu bar
- Left-click the AC Analysis tab and enter the simulation parameters below:

Type of Sweep: Decade
Number of points per decade: 150
Start Frequency: 0.1
Stop Frequency: 1e6 or 1Meg

A decade frequency sweep uses a logarithmic point spacing over each 10:1 frequency decade. Choosing around 101 points per decade, the traced plot appears smooth and a point occurs on each decade boundary.

iii. Run the .ac simulation. Left-click on the output node voltage Vout to probe and plot an output magnitude and phase variations versus frequency as shown in Fig. 9.8. The simulation analysis results in a Bode plot displaying an amplifier gain magnitude versus signal frequency because the input AC is assigned a 1 V amplitude. Thus, a plot of the output signal magnitude is the same as the gain magnitude plot.

(**Note:** The phase trace can be deleted. Also, by right-clicking on a right-vertical axis (phase axis) on the right-hand side and left-clicking the OK, the deleted phase plot can be brought back into the waveform viewer window.)

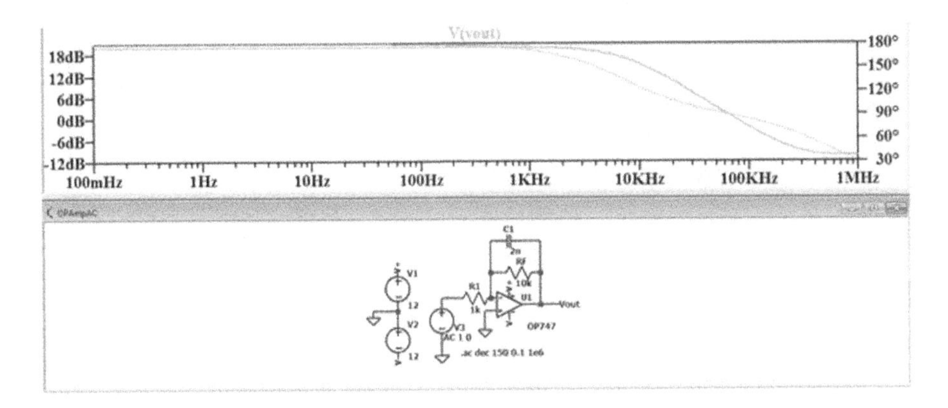

FIGURE 9.8 The Magnitude Response (Bode Plot) of an Active Low-Pass Filter.

iv. Using the numbered cursors, find the circuit's –3 dB frequency where the gain magnitude is 3 dB lower than the maximum at 0 Hz. Choose to attach the Cursor 1st & 2nd to place both the cursors on the plot and measure the circuit –3 dB cut-off frequency as follows:
- Right-click on the trace name at the top of the plot and the Expression Editor appears. In the Attached Cursor menu, select the 1st & 2nd. Left-click the OK.
- Move a mouse around the plot until a yellow 1 (the Cursor 1) appears. Left-click, hold and drag the cursor down to 0.1 Hz. The gain magnitude at 0.1 Hz is very close to the maximum gain that occurs at 0 Hz.
- Move a mouse around the plot until the yellow 2 appears. Left-click, hold and drag a cursor until the Ratio (Cursor2/Cursor1) magnitude display reads –3.00 ± 0.02 dB. Read the Cursor 2 frequency in the Freq box.

The cut-off frequency comes out to be equal to 7.12265 kHz.

Thus, the experimental result almost matches the manual result (Eq. 9.1).

v. Place a marker at the –3 dB point as follows:
   i. In the active plot pane, from the plot menu, select the Plot Settings -> Notes and Annotations -> Label Curs. Pos (see Fig. 9.9). Also, the Draw -> Arrow options can be used to make an arrow from a text label placed manually to a point on the trace. Right-click on the text label to change or delete it.

vi. An AC Analysis can also be done at a single frequency. Right-click on the empty .ac statement placed on the schematic to open the statement editor window. To obtain the simulation result at a single frequency, say π rad/s, in the .ac analysis statement editor, the user should select the **List** option in the pull-down menu next to the Type of sweep field and enter the desired frequency value π rad/s encased in curly braces i.e. {pi/(2*pi)} in the 1st frequency box Type of sweep: List 1st frequency: {pi/(2*pi)}. Now place the complete .ac directive

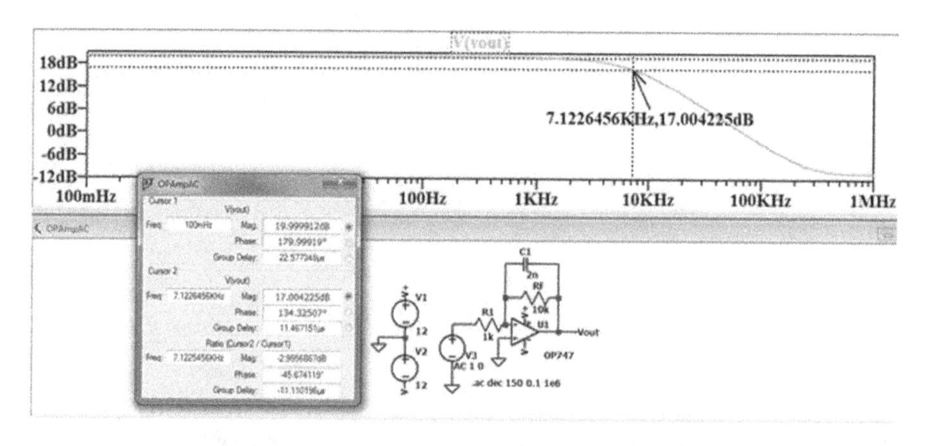

FIGURE 9.9 Reading the Cut-Off Frequency of the Designed Filter.

command on the schematic as shown in Fig. 9.10. Curly braces are used so that the simulator evaluates the expression before running the simulation. The frequency is divided by 2π because LTspice XVII assumes frequencies in Hertz and not in rad/sec. Place the .ac list {pi/(2*pi)} statement shown in a text area at the bottom of the **Edit Simulation Command** window on the schematic by clicking the OK. Press the Run button to obtain the results at the frequency of π rad/s. The output of an AC Analysis at a single frequency of 0.5 Hz in a separate window is shown in Fig. 9.10.

vii. Now, in the 1st frequency box enter a single frequency as an expression containing a global parameter and encased in curly braces e.g. here enter {w/(2*pi)} as shown in Fig. 9.11. Now, the angular frequency parameter w is defined using the .param directive statement. Place the .param w 100 statement (or .param w = 100) on the schematic. To obtain the results at a single frequency of 100 rad/s, press the Run button. The output of an AC Analysis at a single frequency of 100 rad/sec using the .param:

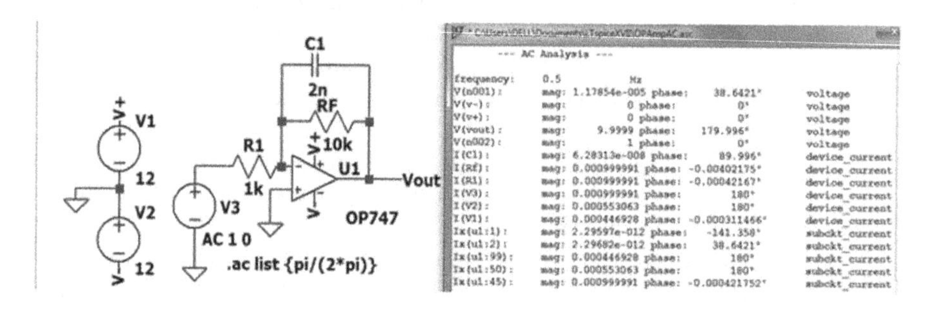

--- AC Analysis ---

frequency:	0.5	Hz			
V(n001):	mag:	1.17854e-005	phase:	38.6421°	voltage
V(v-):	mag:	0	phase:	0°	voltage
V(v+):	mag:	0	phase:	0°	voltage
V(vout):	mag:	9.9999	phase:	179.996°	voltage
V(n002):	mag:	1	phase:	0°	voltage
I(C1):	mag:	6.28313e-008	phase:	89.996°	device_current
I(Rf):	mag:	0.000999991	phase:	-0.00402175°	device_current
I(R1):	mag:	0.000999991	phase:	-0.00042167°	device_current
I(V3):	mag:	0.000999991	phase:	180°	device_current
I(V2):	mag:	0.000553063	phase:	180°	device_current
I(V1):	mag:	0.000446928	phase:	-0.000311466°	device_current
Ix(u1:1):	mag:	2.29597e-012	phase:	-141.358°	subckt_current
Ix(u1:2):	mag:	2.29682e-012	phase:	38.6421°	subckt_current
Ix(u1:99):	mag:	0.000446928	phase:	180°	subckt_current
Ix(u1:50):	mag:	0.000553063	phase:	180°	subckt_current
Ix(u1:45):	mag:	0.000999991	phase:	-0.000421752°	subckt_current

FIGURE 9.10 An AC Analysis at a Single Frequency.

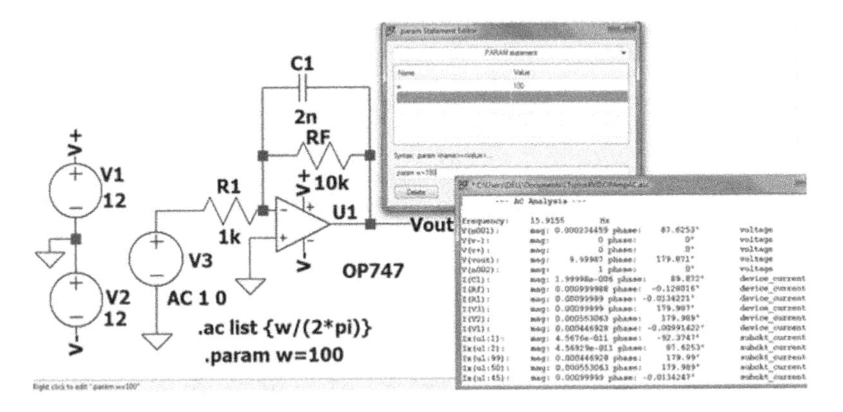

FIGURE 9.11 An AC Analysis at a Single Frequency using the .param.

9.4 PASSIVE FILTER

i. Construct the RC low-pass filter circuit schematic as shown in Fig. 9.12.

The solid lines represent a gain magnitude (in dB unit). The simulated first-order low-pass filter has a cut-off frequency (also called a – 3 dB frequency at which the circuit gain magnitude is –3 dB below its maximum value) equal to 1/(2πRC) = 159 Hz at a 1 kΩ resistor and 1 μF capacitor. From Fig. 9.12, it can be stated that the experimental result is similar to the theoretical result.

9.5 VCVS AS NON-INVERTING OP-AMP

i. Draw the circuit schematic and make the connections so that the VCVS model E1 with the symbol **e** behaves as a non-inverting amplifier as shown in Fig. 9.13.

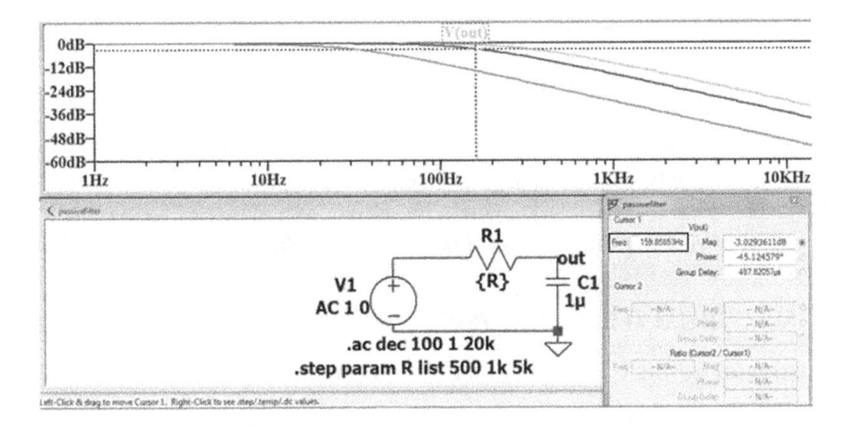

FIGURE 9.12 An AC Analysis with Stepping the Resistance of the RC Passive Filter.

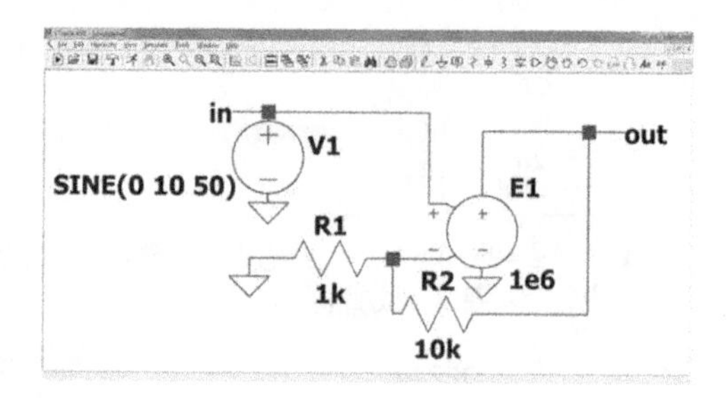

FIGURE 9.13 The VCVS as a Non-Inverting OP-Amp.

Change the independent voltage source V1 to a sine wave of peak amplitude 10 V and frequency 50 Hz. By clicking on the Advanced button in the independent voltage source attribute editor box, different source values for different simulation types can be accessed, such as:

- The Functions setting allows waveform types for transient (i.e. time-domain) simulations
- The (none) Function plus the DC Value setting is for operating point and transient simulations
- The Small signal AC analysis setting is for doing an AC analysis (i.e. frequency-domain or AC sweep) simulations

ii. Open the Select Component Symbol window and type a character **e** in a box to obtain the VCVS component with polarities as shown in Fig. 9.13. Change the VCVS scale factor by right-clicking on the placeholder letter E and entering 1e6 (which stands for 1×10^6). A scale factor does not necessarily need to have a constant value, an equation involving other circuit parameters can also be entered as a scale factor.

iii. Set-up a transient simulation to run for three sine wave cycles by specifying the Stop Time to be equal to 60 ms. Specifying the Maximum Timestep is non-compulsory.

The Output (Fig. 9.14):

(**Note:** Right-click in the middle of the plot window already plotted, and then select the Add Traces option to plot both the waveforms in different window panes.)

Remember: Right-click on an empty area of the plot window, then select the View -> Copy bitmap to Clipboard options to insert (copy and paste) both the traced waveforms in an MS-Word document. Before copying a traced plot to a clipboard for pasting it in a document, it is vital to change a black (the default) waveform background to white.

9.6 SPECTRUM (FFT) ANALYSIS

Simulation of a Single Rectangular Pulse Spectrum:

The information about frequencies (harmonics with frequency values (2×, 3×, 4× ….) integer multiples of the fundamental frequency) present in a signal can be

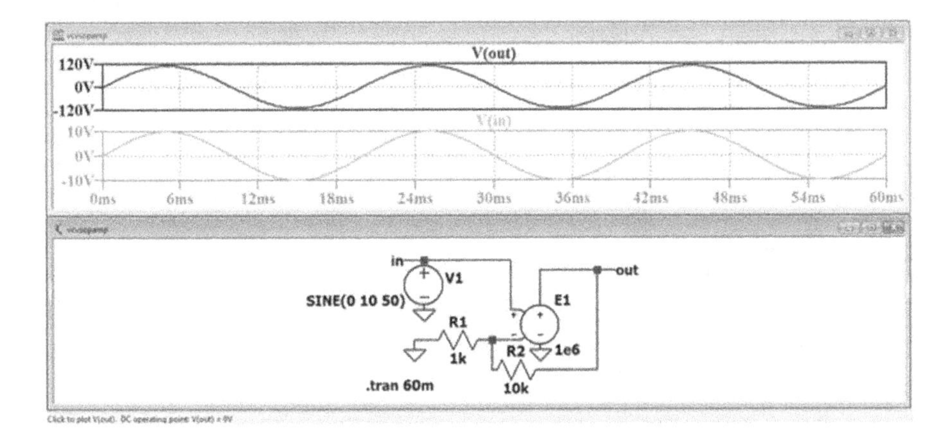

FIGURE 9.14 The VCVS Amplified Output In-Phase with the Input.

obtained using Fast Fourier Transform (FFT). While performing FFT, a minimum sample frequency = 1/(maximum time step) must equal to twice the maximum frequency value in a simulated signal to avoid undesired aliasing effects.

A maximum frequency and dynamic range of FFT can be varied by a varying number of samples used in FFT. The number of samples is given by:

The true number of samples = (simulation time)/(maximum time step)

For example, if a simulation final time is chosen to be 10 ms (start time is 0 ms by default) and a maximum time step value is selected to be 100 ns, number of real samples of a simulated curve is equal to 10 ms/100 ns = 100000 samples.

i. Now draw the circuit diagram for generating a Single Rectangular Pulse across a resistor. Set the properties of the voltage source component (the voltage sources can produce many test signals, whereas selecting the **PWL** function can construct any desired signal). Therefore, right-click on the voltage source component and select the **PWL** function. Configure the PWL voltage source (see Figs. 9.15 and 9.16) to manually create a pulse waveform using the following data:

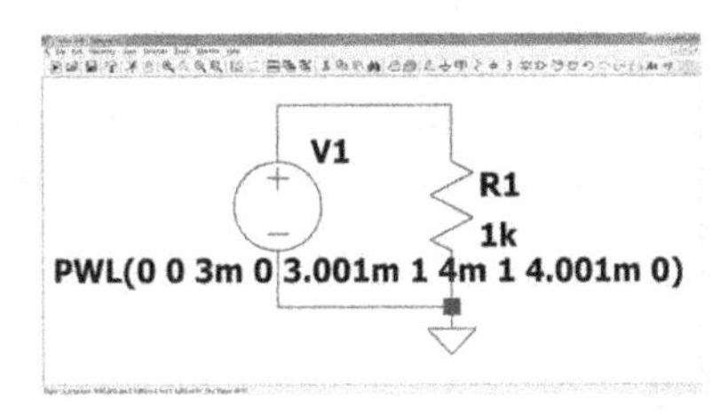

FIGURE 9.15 The Circuit Schematic of a Single Rectangular Pulse.

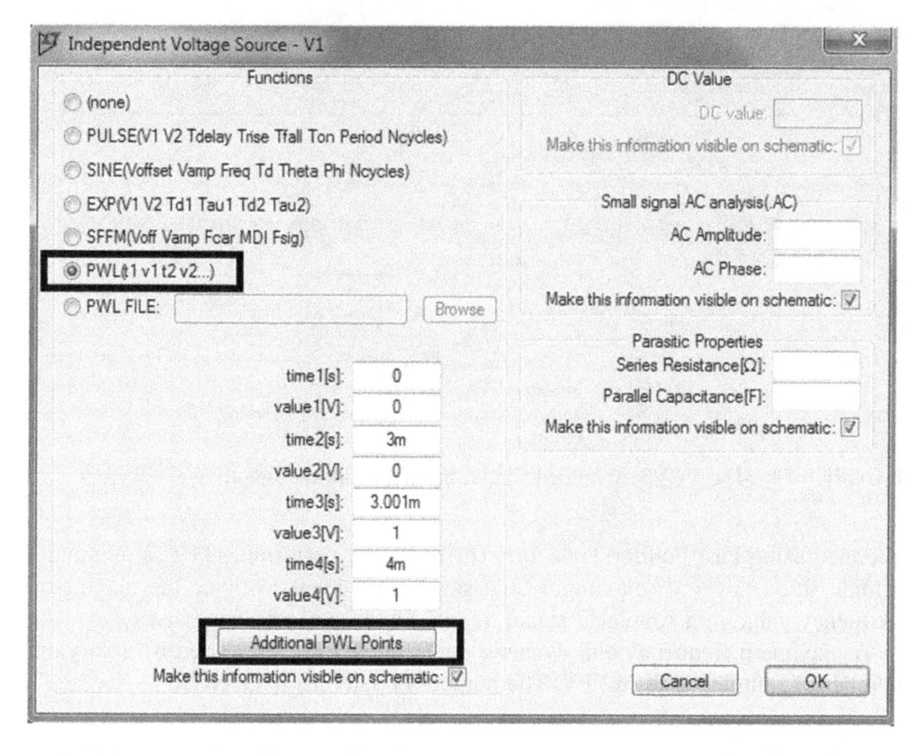

FIGURE 9.16 The Independent Voltage Source – V1 Dialog Window.

0 Volt at t = 0
0 Volt at t = 3 ms
1 Volt at t = 3.001 ms
1 Volt at t = 4 ms
0 Volt at t = 4.001 ms

(**Note:** If the difference between the time values is very less, the signal rises/decays abruptly, otherwise the rise/decay is linear.)

In the given circuit, choose additional points in a generated pulse signal for simulation by clicking on the **Additional PWL Points** (see Fig. 9.16) button.

Add one more data point by giving a time, amplitude value pair as shown in Fig. 9.17.

Thus, selecting the PWL (Piece-Wise Linear) function for the source (voltage or current) component, a complex waveform can be created that consists of straight line segments connecting as many points defined by the user. The syntax contains required parameters as a list of arguments entered as two-dimensional points representing time and voltage (or current) value data pairs. The number of data pairs can be any with time values in ascending order and intervals between time values can also be non-regular.

- The syntax of the PWL function is:
  PWL(0 0 3m 0 3.001m 1 4m 1 4.001m 0) (as the difference between the time values is very less, so abrupt rise/decay)

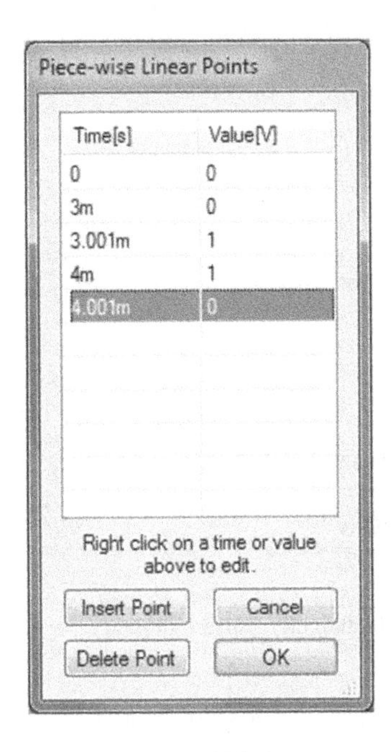

FIGURE 9.17 The Piece-wise Linear Points Window.

- In another form, the time values can be defined relative to the previous time value by prefixing the time value with a + sign as follows:

PWL(0 0 +1m 1 +1m 1 +1m 0)

To repeat (data pairs a specified number of cycles or forever) the PWL source, right-click on the PWL text string placed on the schematic and use the repeat command e.g. **PWL repeat for 3 (time-value pairs) endrepeat** to create three pulses (or three cycles of an input wave).

ii. Choose a transient analysis simulation time of 10 ms (Fig. 9.18) for desired FFT simulation (i.e. desired frequency resolution = 1/simulation time = 1/10 ms = 100 Hz) and use a preset value of number of data samples (Fig. 9.19).
iii. Now, right-click in the plot pane on the waveform viewer window, go to the View -> FFT (Fig. 9.18) to start FFT simulation for 500 points using a preset value of several data samples (Fig. 9.19).

After clicking the OK, a magnitude spectrum (in dB unit) of the generated rectangular pulse is plotted (Fig. 9.20).

iv. Now, modify the magnitude spectrum as follows:
   a. Move a cursor to the scaling of a vertical axis and left-click when a ruler appears to change the default dB magnitude scale to a linear presentation.
   b. Similarly, set a horizontal axis (logarithmic by default) to linear (uncheck the Logarithmic box), enter a start frequency of 0 Hz, tick

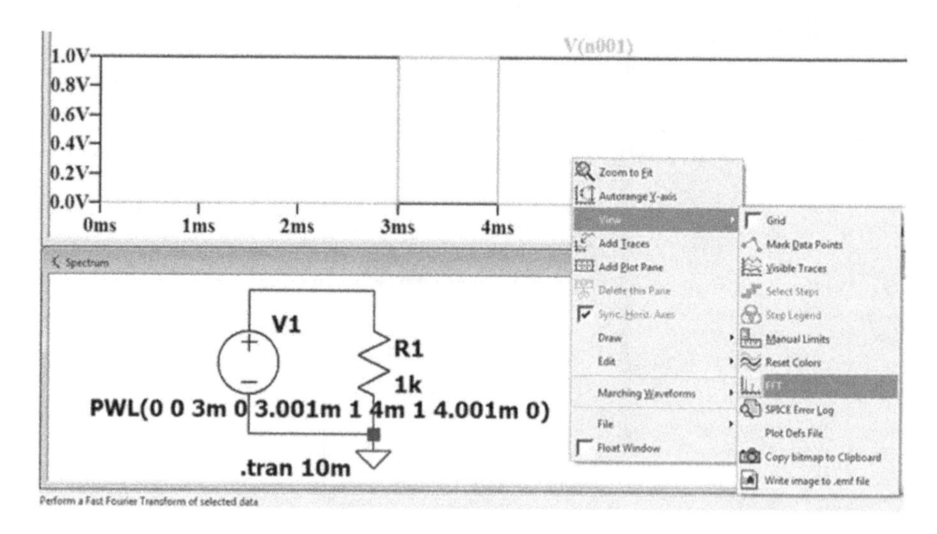

FIGURE 9.18 The Single Rectangular Pulse Output.

frequency of 1 kHz, and a stop frequency of 10 kHz (Fig. 9.21). The diagram starts with a frequency of 100 Hz = 1/10 ms = 1/simulation time (Fig. 9.22).

The output (magnitude spectrum) on a linear scale (Fig. 9.22):

By right-clicking on a right-vertical axis on the right side of the waveform viewer window, a phase spectrum can also be plotted.

c. Finally, choose the Tile Horizontally option under the Window schematic menu to obtain the changed layout of the results as shown in Fig. 9.23:

A single pulse is present due to a transient simulation. An FFT simulation result confirms that the spectrum of non-periodic signals is continuous and there are no discrete frequency lines so that the resulting diagram follows a **Sin x/x** (zeros at every multiple of 1/pulse length and shortening a pulse length increases the frequency of all respective zeros) envelope function.

9.7 NOISE ANALYSIS (the .noise)

By performing a Noise analysis, the noise inherent in a circuit as well as added externally from an outside source can be viewed and a noise voltage density (V/√Hz) for 1Hz bandwidth can easily be plotted for an output noise, input noise, or for any noisy component like a resistor, diode or transistor. SPICE performs a Noise analysis together with an AC analysis statement.

LTspice XVII finds noise sources within individual components of the drafted circuit. A noise analysis involves tracking individual noise sources and plots a total output noise after adding the noises from these sources together.

Select Waveforms to include in FFT

NOTE: Fourier components are normalized to correspond to the time domain RMS amplitude.

Ctrl-Click to toggle

OK

Cancel

V(n001)
I(R1)
I(V1)

Number of data point samples in time: 262144

Quadratic interpolate uncompressed data

Time range to include

Start Time: 0s

End Time: 10ms

Use Extent of Simulation Data

Use current zoom Extent

Specify a time range

Binomial Smoothing done before FFT and windowing

Number of Points: 500

Windowing (Periodic and normalized to unit area)

Windowing Function: (none)

Kaiser-Bessel Parameter Beta: 2

Preview Window

NOTE: The DC component is removed before windowing.

Reset to Default Values

FIGURE 9.19 The FFT Simulation.

Resistor noise is modeled as a noise voltage in series with a resistor.

ii. **Simulating Spectral Noise Density in a Resistor**

A voltage spectral density is simply a square root of power spectral density (power generated by a resistor in a 1 Hz bandwidth):

$$N = (4\ k\ T\ R)^{½} \quad (V/\sqrt{Hz}) \tag{9.2}$$

where N = Noise Voltage Spectral Density and k = Boltzmann's Constant (1.38×10^{-23} J/K)

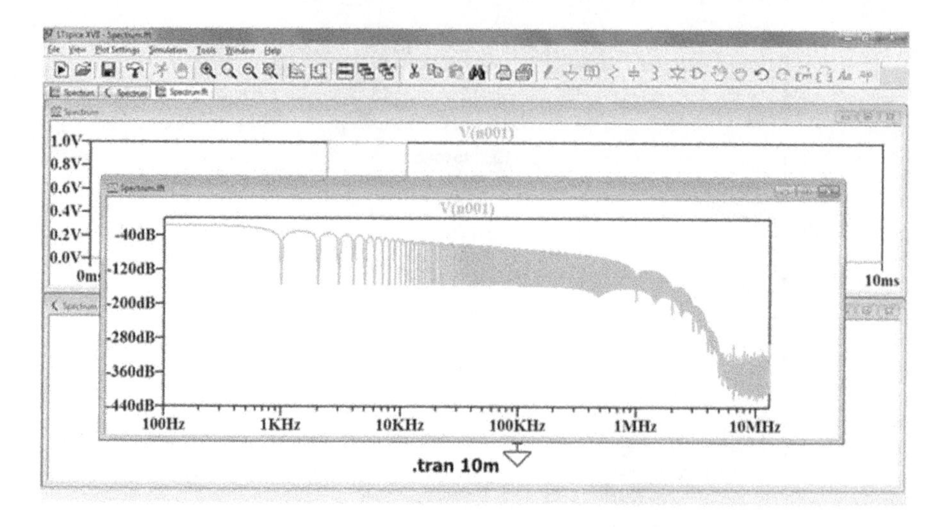

FIGURE 9.20 The Magnitude Spectrum of the Rectangular Pulse.

Horizontal Axis
Quantity Plotted: frequency
Eye Diagram
Axis Limits
Left: 0Hz
tick: 1KHz
Right: 10KHz
Logarithmic
Cancel
OK

FIGURE 9.21 The Horizontal Axis Dialog Window.

T = Temperature in Kelvin (by default LTspice XVII uses a room temperature i.e. 27 °C)

R = Resistance of a resistor in Ω

(**Note:** An ideal resistor contributes a voltage noise $v_n = \sqrt{4kTBR}$, where B is bandwidth in Hz)

A resistor produces a thermal noise considered as approximately **white** in case of an ideal **resistor which implies** that a power spectral density is nearly constant throughout a frequency spectrum. A noise voltage spectral density is the same at all frequencies in a resistor. An amount of noise depends on a value of resistance, bandwidth and temperature.

i. Draw the circuit diagram of a voltage divider (generally used to establish a voltage reference or bias in a circuit with a single supply voltage). In the Edit Simulation Command window, select the Noise tab and fill in the parameters as shown in Fig. 9.24 to build the .noise analysis command.

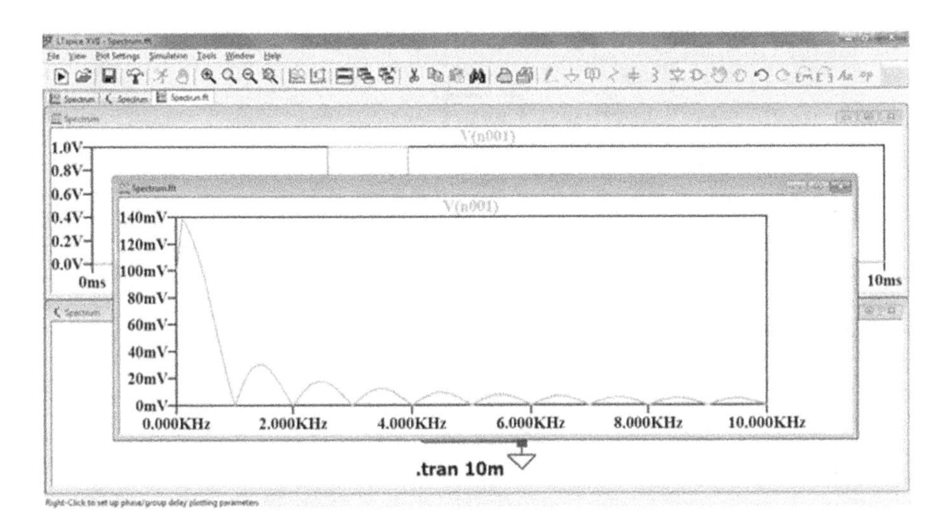

FIGURE 9.22 The Magnitude Spectrum on a Linear Scale.

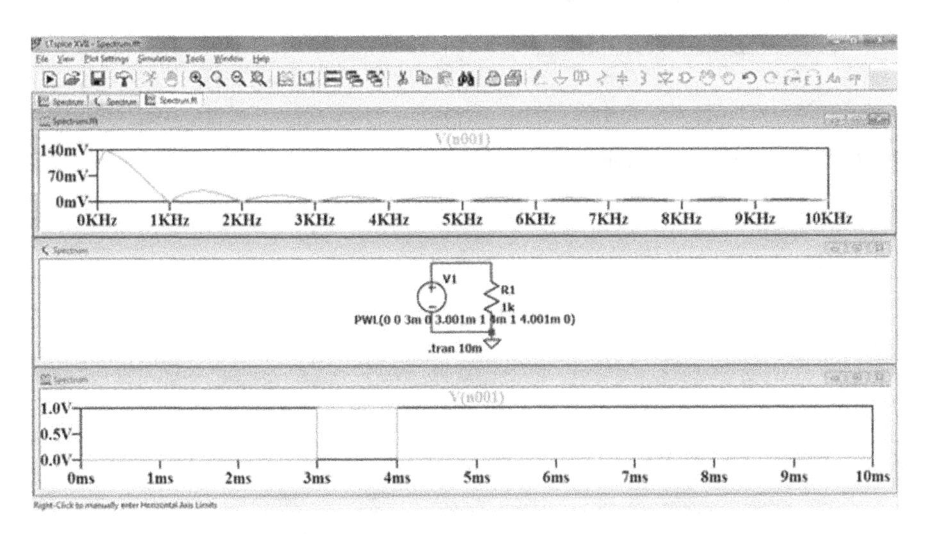

FIGURE 9.23 The Generated Rectangular Pulse and its Magnitude Spectrum.

ii. Click the OK and place the .noise analysis complete statement (.noise V(Ref) V1 dec 100 2 10k) on the schematic. Run the simulation. Click on the **Ref** node (a name of the output plot automatically changes to **V(onoise)** by the simulator) when the probe appears (Fig. 9.25).

The plot shows a flat line at 2.35 nV/Hz^{½} (a total noise at the output due to all individual noise sources added together).

iii. Click on the body of the R1 to add a plot of noise at the output coming from only the R1.

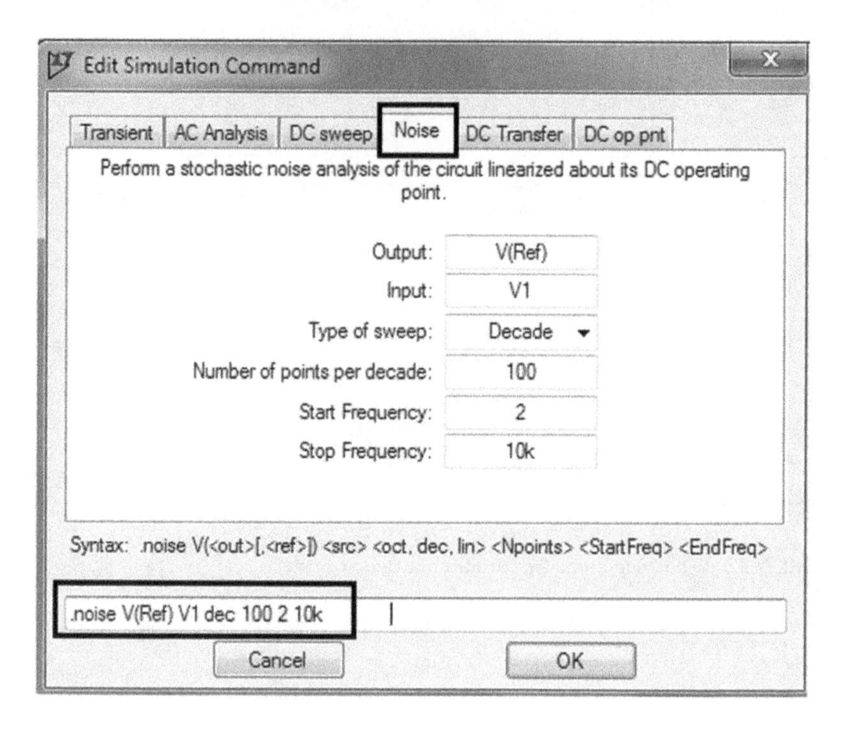

FIGURE 9.24 The Noise Tab Parameters.

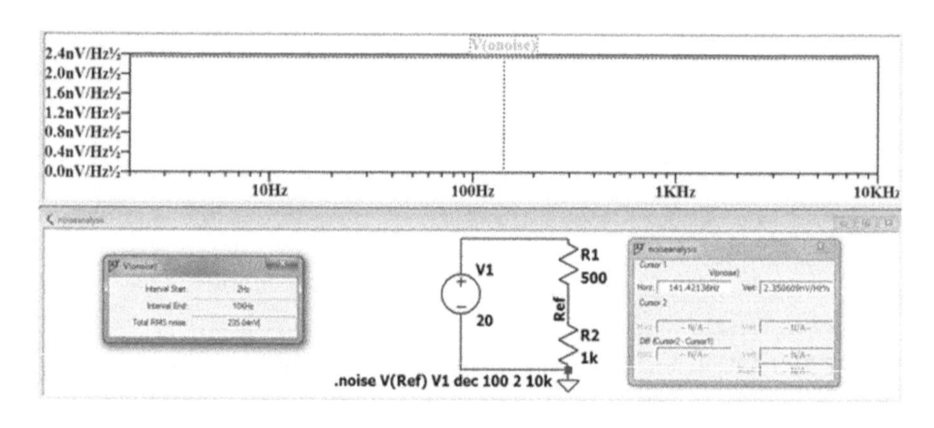

FIGURE 9.25 The Simulated Output using the .noise Analysis.

iv. Noise can be integrated over a selected bandwidth to display a total RMS noise by using a Ctrl + click on the datatrace label V(onoise). Also, a total noise over a limited bandwidth rather than the bandwidth specified in the .noise directive can be computed by modifying the left (low) and right (high) frequency limits of the waveform viewer by clicking on a horizontal axis.

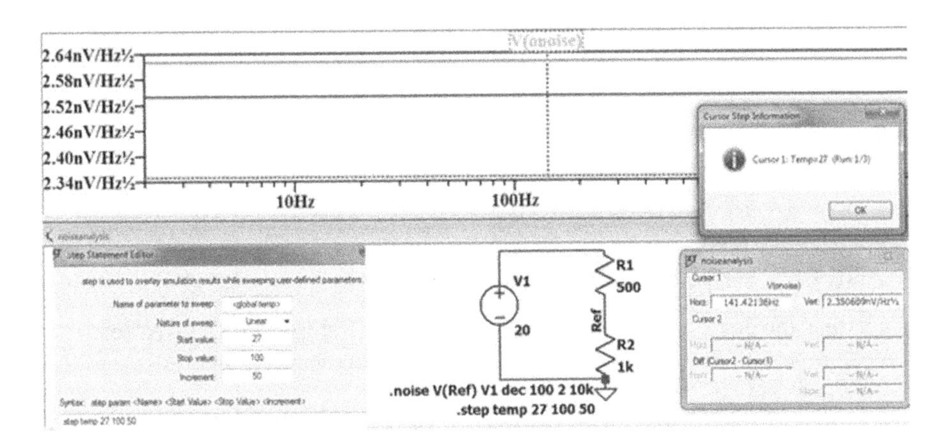

FIGURE 9.26 The Simulated Output for a Range of Temperatures.

v. Input noise can also be plotted by right-clicking on an empty area in the plot pane and selecting the View -> Visible Traces options to open the Select Visible Waveforms window containing a list of waveforms that can be plotted.

9.7.1 Noise over a Temperature Range

By default, LTspice XVII employs a temperature of 27°C but noise modeling can be easily done at different temperatures using the .step directive (to specify a range in steps, see Fig. 9.26) or .options directive (to specify a single temperature e.g. .options temp = 100). The simulator provides a variable **temp** (the keyword) as the built-in function (Fig. 9.26).

(**Note:** The attached numbered cursor can be easily navigated from a dataset to another dataset with an up/down keyboard cursor key. Here, a down key is used because a plot for the start value is the bottommost.)

9.8 LOGIC GATES SIMULATION

- **Simulating an AND gate:**

    i. Place the AND gate component on the schematic (Fig. 9.27).

The basic AND gate component (other gates are also available, see LTspice XVII Help) has 5 inputs (1 ... 5) on the left, one common return pin named 8 at the bottom plus 2 output pins (one inverting and the other non-inverting). Thus, the gate component can be used both as an AND gate as well as a NAND gate.

Any unused inputs and/or outputs should be connected to the **bottom** pin **8** so that the simulator distinguishes them as unused and removes them from a simulation. Also, the pin 8 (along with the unused I/Os) should be connected to a common ground (GND, the global node 0). No external power supply is used for

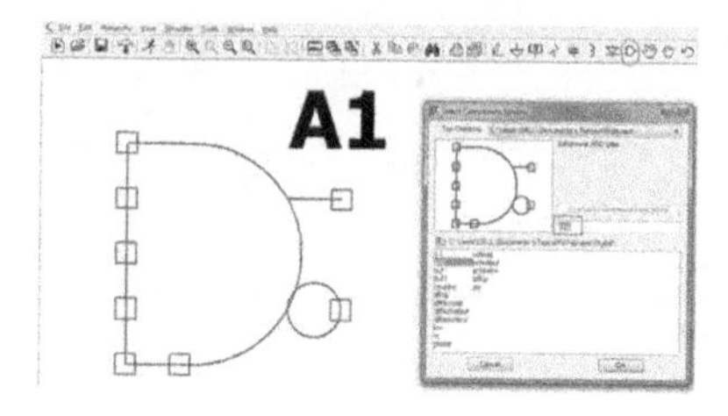

FIGURE 9.27 Simulating an AND Gate.

simulating the logic gates (other than the common return pin 8). It is better to provide a path to the ground for the used output by connecting the resistor to the ground to avoid an error.

ii. Draw the circuit schematic as shown in Fig. 9.28.
iii. Run the .op simulation analysis (with 2 DC input values) to obtain the following output (Fig. 9.29):
iv. Now, label the input nodes as in1 and in2. Apply symmetric pulses (minimum value = 0V and maximum value = + 1 V) to the 2 inputs for simulating AND gate so that the input 1 is a symmetric pulse signal having a rise time = fall time = 0.1 ns and pulse length = 1 ms with a time period = 2 ms and the other input 2 has a rise time = fall time = 0.01 ns and pulse length = 2 ms with a time period = 4 ms. Right-click on the voltage source V1 and select the PULSE function and do the entries for an input 1 as follows (Fig. 9.30):

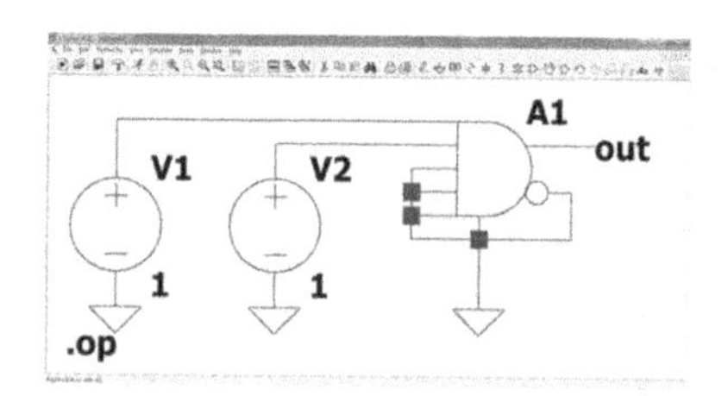

FIGURE 9.28 The Circuit Schematic for an AND Gate Simulation.

```
* C:\Users\DELL\Documents\LTspiceXVII\ANDgate.asc

       --- Operating Point ---

V(n001):        1            voltage
V(n002):        1            voltage
V(out):         1            voltage
I(V2):          0            device_current
I(V1):          0            device_current
I8(A1):         -0           device_current
I7(A1):         0            device_current
```

FIGURE 9.29 The Simulated Output of the AND Logic Gate.

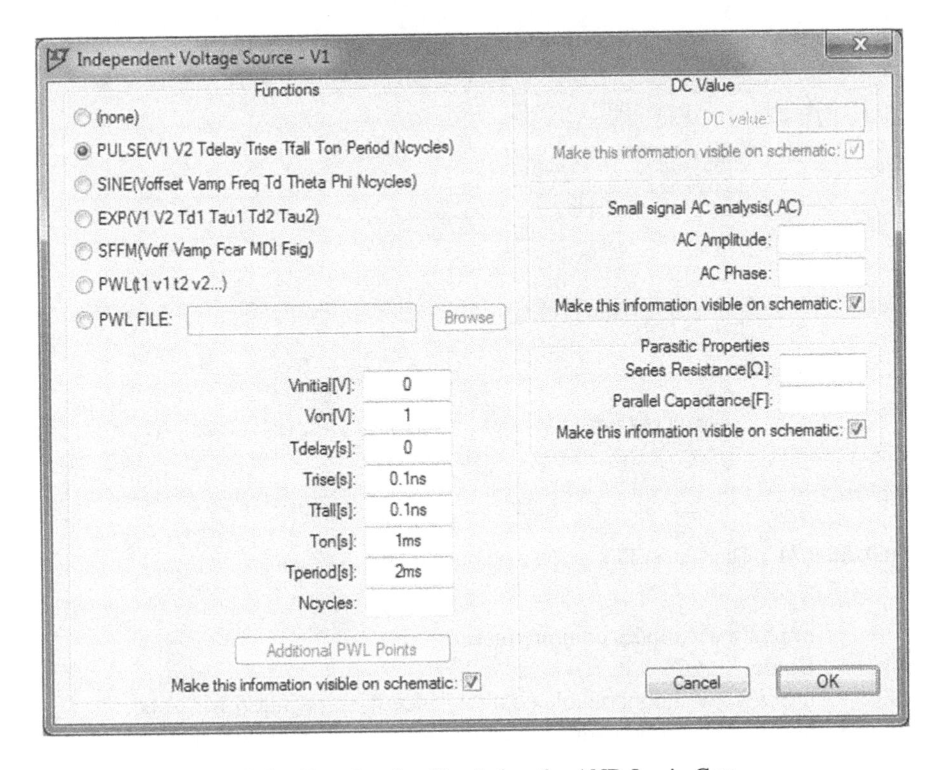

FIGURE 9.30 The Pulse Function for Simulating the AND Logic Gate.

v. Similarly, right-click on the voltage source V2 and do the entries for an input 2.
vi. Re-run a transient analysis simulation for 6 ms. Probe the in1, in2, and out nodes to obtain the following output (Fig. 9.31):

9.9 INSERTING AN EXTERNAL SPICE FILE

To use a new model for the new element or device similar to the one that exists in the simulator library while doing a simulation, add an external SPICE model directly into the drawing as a text file so as to pass specific parameters to the existing model by copying and pasting it in the directory where the drafted circuit schematic is saved. Now, place the .include directive statement (a piece of text that starts with .include followed by the name of the text file containing the model to be added) on the schematic to take into account the text file into a simulation. Thus, a simple third party SPICE model can be imported into the simulator software following the steps:

- Add the generic component to the schematic that represents the symbol of the SPICE model.
- Download the SPICE model file and insert it into the same directory where the drafted circuit diagram is saved for simulation. For example, to look for SPICE new models for the new components, the user can use the URLs:

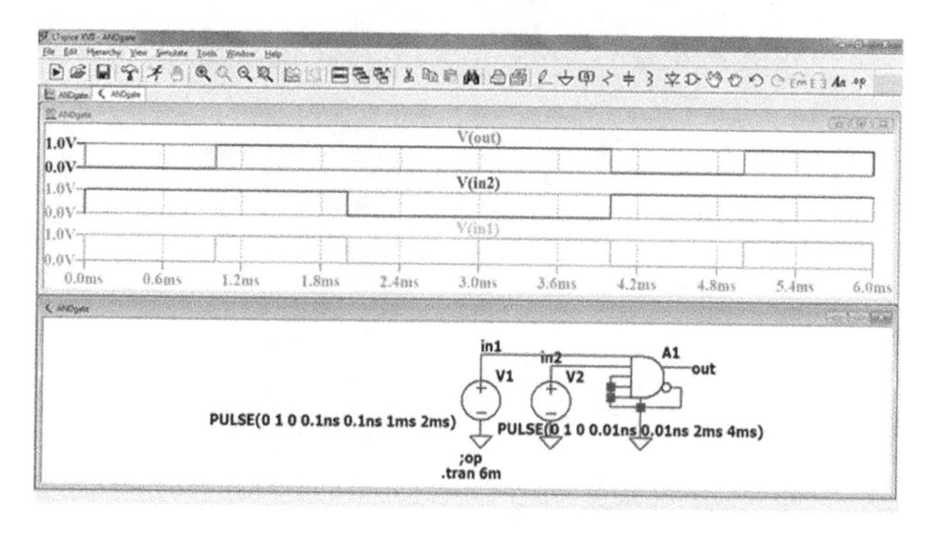

FIGURE 9.31 The Output Plot of the AND Logic Gate.

- https://www.diodes.com/products/discrete,
- Diode Test Circuit,
- http://www.simonbramble.co.uk/lt_spice/lt_spice_files/LM324.txt,
- http://www.simonbramble.co.uk/lt_spice/lt_spice_files/DI_BAT54 .txt, etc.

- Open the downloaded SPICE model file (text document notepad file) and notice the **filename** of the SPICE model along with the file **extension** (type of file). To avoid spelling mistakes, use Windows Explorer (or Chrome) to highlight and copy the filename including the filename extension (.txt). Paste the text after the include statement.
- Add the .include SPICE directive to the drafted circuit schematic in order to use the added model.
- Right-click on the .include directive and place the name of the text file containing the new model along with the filename extension (.txt) immediately after the .include directive to complete the statement.
- Open the SPICE model file imported directly into the drawing and note the name of the model (the text immediately after the .model directive and before the part designator.
- Press a Ctrl key, and then right-click over the generic component to change the Value field parameter to match the model name.
- Ensure the .include SPICE directive contains the exact filename of the SPICE model including the filename extension (.txt).
- Ensure the name of the generic component exactly matches the SPICE model name.

An example of adding a new model:

i. Below is an example showing that a new Zener diode model can be added to the circuit by using the link as follows:
https://www.diodes.com/products/discrete/diodes-and-rectifiers/diodes/zener-diodes/
ii. Place the standard Zener diode from the LTspice XVII library and draft the circuit diagram in which the voltage source component is connected across a series combination of the resistor model and standard Zener model. Set the voltage source to be DC having a value of 5 V and the resistance of the resistor component to be 1 kiloohm . Save the Drafted circuit.
iii. Download a SPICE model 1SMB5928B (.txt file) of a Zener diode and save it in the same directory as the drafted circuit.
iv. Add the following SPICE directive to the drafted circuit schematic using the .op schematic tool .op symbol:

.include 1SMB5928B.spice.txt

v. To avoid spelling mistakes, open the downloaded file (say Notepad document) and use the File -> Save as (or use Windows Explorer) to highlight and copy the filename including the filename extension (.txt). Paste the text after the .include statement.
vi. From the SPICE model file, note the name of the model (here, it is 1SMB5929B) and copy it. Afterward, press a Ctrl + right-click over the standard Zener diode symbol and paste the text 1SMB5929B into the Value field. Comments should not be added in the SPICE Model field. The final circuit looks like as shown in Fig. 9.32:
vi. Run a transient simulation for 60 ms.

If errors are generated, re-check whether the SPICE file (type a name of the file, say here 1SMB5928B.spice.txt in the File name box to search for it) is loaded into the same directory as the simulation file named external or the component parameter value name is the same as is specified in the SPICE model or the filename and its type (specified by giving an extension) in the .include statement exactly matches the

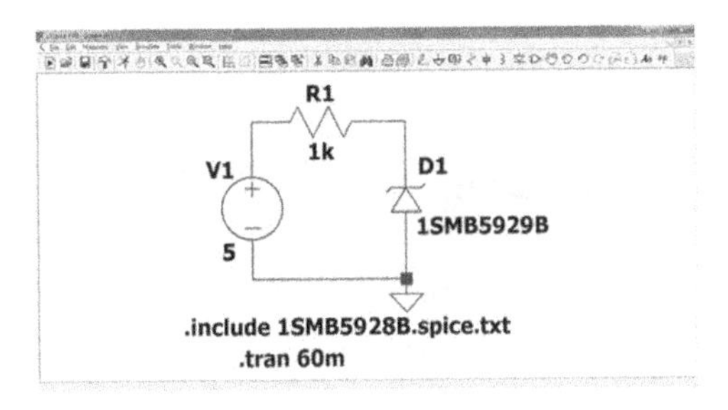

FIGURE 9.32 The Circuit Schematic.

SPICE model filename and its type. Also, the loaded SPICE model file may be opened within LTspice XVII for checking the presence of unwanted formatting characters in the contents.

The Output (Fig. 9.33):

A Zener diode behaves like a normal ordinary diode when forward biased.

vii. Now, set the voltage source to be sinusoidal having a peak amplitude of 20 V and a frequency of 50 Hz (Fig. 9.34).

viii. Run a transient simulation for 60 milliseconds to obtain the output (Fig. 9.35):

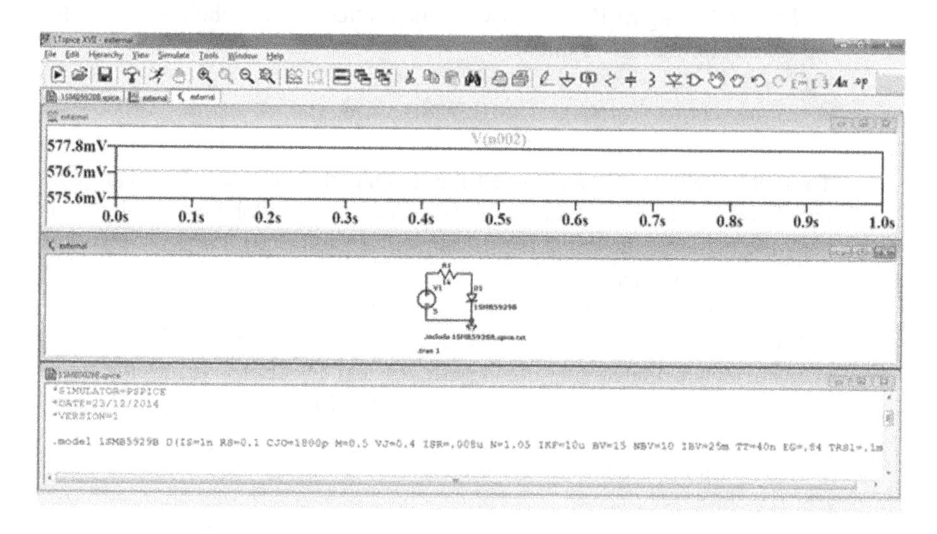

FIGURE 9.33 The Simulated Output.

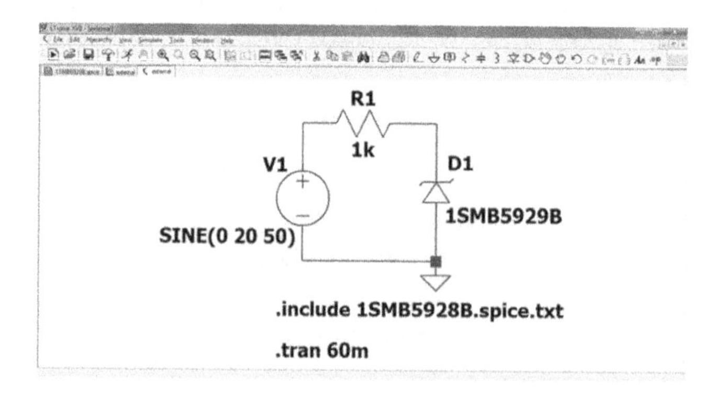

FIGURE 9.34 The Circuit Schematic.

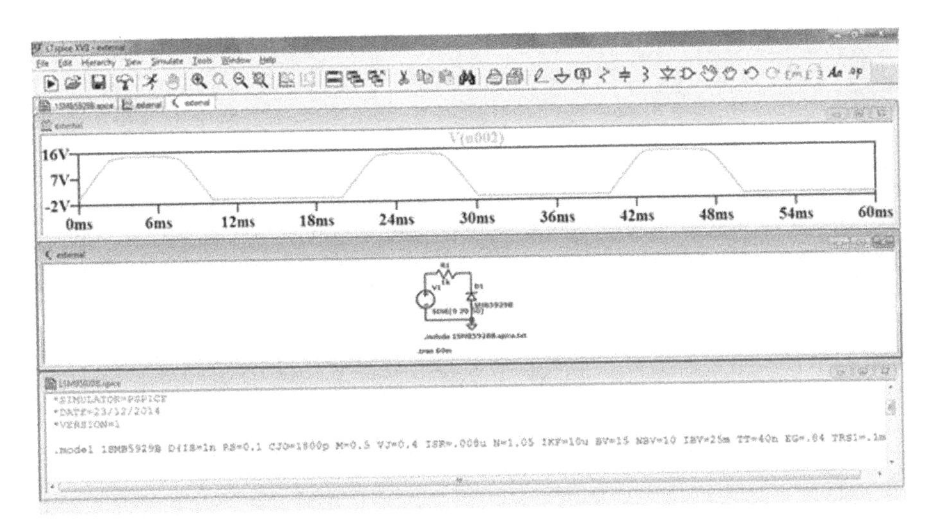

FIGURE 9.35 The Simulated Output.

For a positive half cycle, the reverse-biased Zener diode remains OFF as long as the input is less than the specified Zener voltage so that the simulated output follows the input. When the input becomes more than the specified Zener voltage (V_Z) during a positive half cycle, the Zener operates in a reverse breakdown region to behave as a battery of V_Z volts and the output becomes equal to the Zener voltage. Whereas, for a negative half cycle, the Zener is forward biased and behaves as an ordinary diode.

9.10 PULSE RESPONSE TO RC CIRCUIT (PWM FILTERING)

When a pulse signal is applied to an RC circuit, an output wave shape depends on a relationship of the circuit time constant (RC) and frequency (Time-Period T) of the input pulses. When a pulse width (ON period) and time between pulses (OFF time) are shorter than five-time constants (or 10RC > T), a capacitor does not completely charge or discharge. Thus, a circuit whose time constant RC (or transient time) is large (usually 10 times greater) as compared to a period of an input pulse behaves as an integrator and produces a ramp up and then a ramp down output to an applied pulse input. If a time constant is further increased, an output may approach a constant level equal to an average of an input.

Whereas, if ten times a circuit time constant RC is small as compared to a period of an input pulse (or 10RC < T), an output waveform across a capacitor closely resembles that of input. If a time constant (rate of charging and discharging of a capacitor) is further reduced or a pulse is made longer, a capacitor stays charged for a longer time and also remains fully discharged longer so that an output voltage shape perfectly matches that of input.

If an input is a sine wave, an RC circuit behaves as a simple low-pass filter (LPF) and does not act as an integrator so that an amplitude of an output wave is reduced along with its phase-shifted relative to an input wave. A low pass filter circuit acts

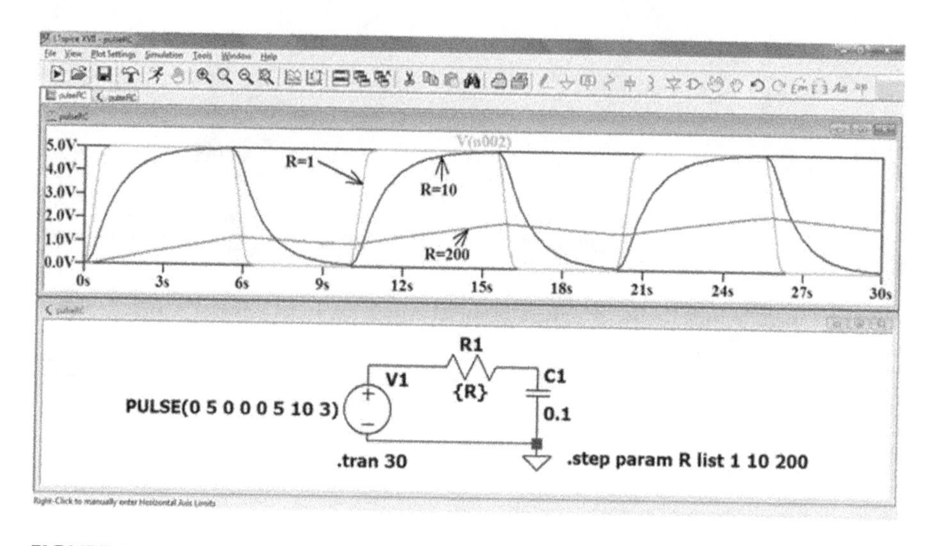

FIGURE 9.36 Sweeping Resistance for a Transient Simulation.

as an integrator only when an input wave is a square wave and its time period is much shorter than a circuit time constant (RC).

The Simulated Output of the Drafted Circuit Shown in Fig. 9.36:

Generate a pulse signal having an initial voltage (off voltage) of 0 V, on-voltage of 5 V, on-period of 5 s and total time period of 10 s by selecting the PULSE function in the V1 component menu window PULSE(V1 V2 Tdelay Trise Tfall Ton Period Ncycles) .

LTspice XVII can also generate an efficiency report. For generating a report, click on the View -> Efficiency Report -> Show on Schematic menus. An efficiency report is generated when there is only one voltage source (as an input) and either one current source or a load called Rload (which is assumed to be the load).

Index